ISAAC TODHUNTER

RESEARCHES

IN THE

CALCULUS OF VARIATIONS

PRINCIPALLY ON THE THEORY OF

DISCONTINUOUS SOLUTIONS

AN ESSAY

TO WHICH THE ADAMS PRIZE WAS AWARDED IN THE UNIVERSITY OF CAMBRIDGE IN 1871

Elibron Classics
www.elibron.com

Elibron Classics series.

ISBN 1-4212-6473-0 (paperback)
ISBN 1-4212-6472-2 (hardcover)

This Elibron Classics Replica Edition is an unabridged facsimile of the edition published in 1871 by Macmillan and Co., London and Cambridge.

RESEARCHES

IN THE

CALCULUS OF VARIATIONS.

Cambridge:
PRINTED BY C. J. CLAY, M.A.
AT THE UNIVERSITY PRESS.

RESEARCHES

IN THE

CALCULUS OF VARIATIONS,

PRINCIPALLY ON THE THEORY OF

DISCONTINUOUS SOLUTIONS:

An Essay

TO WHICH THE ADAMS PRIZE WAS AWARDED IN THE UNIVERSITY OF CAMBRIDGE IN 1871.

BY

I. TODHUNTER, M.A. F.R.S.

LATE FELLOW AND PRINCIPAL MATHEMATICAL LECTURER OF ST JOHN'S COLLEGE, CAMBRIDGE.

London and Cambridge:
MACMILLAN AND CO.
1871.

PREFACE.

THE subject of this Essay was prescribed in the following terms by the Examiners:

THE ADAMS PRIZE.

A determination of the circumstances under which discontinuity of any kind presents itself in the solution of a problem of maximum or minimum in the Calculus of Variations, and applications to particular instances.

It is expected that the discussion of the instances should be exemplified as far as possible geometrically, and that attention be especially directed to cases of real or supposed failure of the Calculus.

E. ATKINSON, *Vice-Chancellor.*
J. CHALLIS.
A. CAYLEY.
J. CLERK MAXWELL.

CLARE COLLEGE LODGE,
April 21, 1869.

It was after much hesitation that I resolved to discuss the subject; the fact that it had given rise to some controversy, however, naturally led me to enforce what I believed to be the correct and adequate explanation of the difficulties which had been raised.

The Essay then is mainly devoted to the consideration of discontinuous solutions: but incidentally various other questions in the Calculus of Variations are examined and I think elucidated. I entertain no doubt of the substantial accuracy of the theory here developed; but at the same time I am aware that in an extensive investigation, which is original and somewhat abstruse, there may be a few subordinate statements which require to be restricted or corrected. I indulge the hope, however, that on the whole I shall have definitely contributed to the extension and the improvement of our knowledge of this refined department of analysis.

The Essay is published as it was originally written; with the exception of the mistakes in algebraical work which occur occasionally in manuscript, but are rendered evident in the clearer form of print. Also a few references have been supplied, and some short remarks, chiefly on passages to which attention had been drawn by the Examiners: these slight additions are enclosed within square brackets.

The laborious task of correcting the press has been undertaken for me by my friend the Rev. J. Sephton; and as on other occasions I am much indebted to him for thus aiding me with his knowledge and accuracy.

I. TODHUNTER.

CAMBRIDGE,
26th October, 1871.

CONTENTS.

CHAPTER I.

PAGE

MAXIMUM OR MINIMUM OF AN INTEGRAL WHICH INVOLVES ONLY ONE DIFFERENTIAL COEFFICIENT 1

CHAPTER II.

THEORETICAL INVESTIGATIONS 13

CHAPTER III.

DISCONTINUITY PRODUCED BY CONDITIONS 32

CHAPTER IV.

MINIMUM SURFACE OF REVOLUTION 54

CHAPTER V.

MAXIMUM SOLID OF REVOLUTION 68

CHAPTER VI.

PROBLEMS ANALOGOUS TO THAT IN CHAPTER V. 100

CHAPTER VII.

BRACHISTOCHRONE UNDER THE ACTION OF GRAVITY . . . 126

CHAPTER VIII.

PROBLEM OF LEAST ACTION 147

CHAPTER IX.

PAGE
Solids of Minimum Resistance 167

CHAPTER X.

Solid of Minimum Resistance with given Volume . . 196

CHAPTER XI.

James Bernoulli's Problem 220

CHAPTER XII.

Multiple Solutions 243

CHAPTER XIII.

Area between a Curve and its Evolute 250

CHAPTER XIV.

Miscellaneous Observations 259

CHAPTER I.

MAXIMUM OR MINIMUM OF AN INTEGRAL WHICH INVOLVES ONLY ONE DIFFERENTIAL COEFFICIENT.

1. In order to arrive at a knowledge of the circumstances under which discontinuity occurs in problems of the Calculus of Variations we shall discuss various problems, beginning with some which are extremely simple. We shall find that speaking generally discontinuity is introduced by virtue of some restriction which we impose, either explicitly or implicitly, in the statement of the problems which we propose to solve.

We do not define the word *discontinuity;* but we shall always make it obvious in what sense we use it as we proceed.

In our investigations we shall never ascribe any variation to the independent variable but only to the dependent variable: in this respect we follow the practice of some of the most eminent authorities on the subject.

[Although a preference is thus expressed for the method of treating the Calculus of Variations which has been adopted by Strauch and Jellett, yet it must not be supposed that this is of importance for the following pages. The results are not affected by this circumstance, although the investigations are rendered in some cases more simple and intelligible than they would otherwise have been.]

2. Let p stand for $\frac{dy}{dx}$; and let $\phi(p)$ denote a given function of p: required the curve for which the integral $\int \phi(p)\,dx$ taken between fixed limits is a maximum or a minimum.

Let $u = \int \phi(p)\,dx$, then, as far as terms of the second order inclusive,

$$\delta u = \int \left\{ \phi'(p)\,\delta p + \phi''(p)\,\frac{(\delta p)^2}{2} \right\} dx$$

$$= \phi'(p)\,\delta y + \int \left\{ -\delta y \frac{d}{dx}\phi'(p) + \phi''(p)\,\frac{(\delta p)^2}{2} \right\} dx.$$

Then we require by the usual theory

$$\frac{d}{dx}\phi'(p) = 0;$$

therefore $$\phi'(p) = \text{constant} \quad\text{.................}\quad (1).$$

The term outside the integral sign in the value of δu vanishes, since the extreme points of the curve are supposed to be fixed.

Thus δu reduces to $\frac{1}{2}\int \phi''(p)\,(\delta p)^2\,dx$.

From (1) we obtain *constant* values of p; for any such value $\phi''(p)$ if it does not vanish will be permanently positive or permanently negative throughout the limits of the integration; thus in the former case we have a minimum value of the integral, and in the latter case a maximum value.

3. We may remark here that it is very important to avoid the common error of using the words *the greatest value* when we ought to restrict ourselves to the words *a maximum value*, and the words *the least value* when we ought to restrict ourselves to the words *a minimum value*. In the present essay we shall use the words *greatest* and *least*, and other similar terms, only when they are strictly applicable.

4. Now return to equation (1). The required curve must be rectilinear; as the extreme points are given the value of p is

known: thus the value of the constant in (1) is determined. Therefore we have apparently only one solution, which furnishes a maximum or a minimum according as $\phi''(p)$ is negative or positive.

But a little consideration shews that it is quite possible to have cases of the problem which require another solution. For instance $\phi''(p)$ may be negative for the known value of p, and yet it may be obvious that there must be some line straight or curved for which the integral has its least value; in fact such a statement must in general be true for any form of the function ϕ.

5. We may naturally ask then if the constant in (1) must necessarily have the same value throughout the limits of integration. Suppose, if possible, that $\phi'(p)$ is equal to C_1 for one part of the required line, and equal to C_2 for the remaining part. Then the integrated term $\phi'(p)\,\delta y$ would introduce into δu the terms $C_1\delta y$ and $C_2\delta y$ *with opposite signs* at the point of the required line corresponding to the change from C_1 to C_2. Thus C_1 and C_2 not being equal δu will not vanish to the first order; and therefore we do not obtain a solution of the problem.

6. But we see from this that there is nothing to prevent us from having a solution made up of *more than one straight line*, corresponding to different values of p found from the equation

$$\phi'(p) = 0 \quad\dots\dots\dots\dots\dots\dots\dots\dots\dots (2).$$

Every condition of the problem may then be satisfied; at least in many cases.

Thus suppose we take two values p_1 and p_2 found from (2), and draw the corresponding straight lines, one through one of the given points, and the other through the other given point. Then if $\phi''(p_1)$ and $\phi''(p_2)$ are both negative we obtain a maximum; and if $\phi''(p_1)$ and $\phi''(p_2)$ are both positive we obtain a minimum. But if $\phi''(p_1)$ and $\phi''(p_2)$ are of opposite signs we do not obtain either a maximum or a minimum. For δu reduces to

$$\frac{1}{2}\phi''(p_1)\int(\delta p)^2 dx + \frac{1}{2}\phi''(p_2)\int(\delta p)^2 dx,$$

where each integral extends between the limits of x which belong to the corresponding value of p. We may suppose δp *to vanish if we please* through one of the two portions into which our integral is divided: thus in this way we can make δu have which sign we please; and therefore with these values of p there is neither a maximum nor a minimum.

[Instead of taking $\phi'(p)=0$ the more general form $\phi'(p)=C$ ought to have been taken; for this may lead to discontinuous solutions by furnishing different values of p. The discontinuity would be like that illustrated in Art. 9.]

Let us now take some particular cases.

7. I. Suppose $\phi(p)=p(1+p^2)$.

Then $$\phi'(p)=1+3p^2, \quad \phi''(p)=6p.$$

Here $\phi'(p)$ cannot vanish. The only solution is the straight line which joins the two given points; and this makes the integral a minimum: for we may suppose that p is positive. Moreover as there must be a *least* value of the integral in this case, it is certain that the minimum value which we have obtained is the *least* value.

No maximum presents itself in this case. In fact we can make $\int \phi(p)\,dx$ as large as we please. Thus in the diagram let A

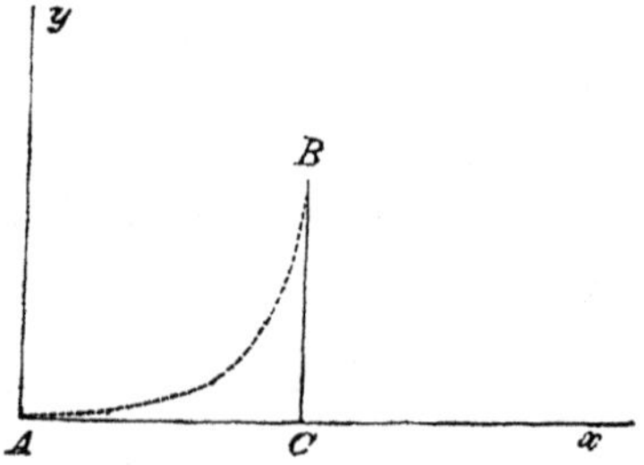

and B be the given points; take A for the origin, and from B draw BC perpendicular to the axis of x. Then $\phi(p)$ vanishes along AC, and is infinite along BC; and we can draw a curve very near to AC and CB for which $\int \phi(p)\,dx$ will be as great as we please. But we do not obtain a *maximum* in the technical sense of that word. We can indeed get a greater result by making our curve go *below* AC.

Difficulties might be suggested as to this case. For as x does *not vary* along CB it might be said that $\int \phi(p)\,dx$ along CB must vanish, since the limits of the integration coincide. If however we transform $\int p(1+p^2)\,dx$ into $\int dy\,(1+p^2)$ we obtain an integral in which the limits do not coincide.

8. II. Suppose $\phi(p) = \dfrac{p}{1+p^2}$, and that the straight line which joins the fixed points makes with the axis of x an angle less than 60°.

Then $\phi'(p) = \dfrac{1-p^2}{(1+p^2)^2}$, $\phi''(p) = -\dfrac{2p(3-p^2)}{(1+p^2)^3}$.

Here the straight line which joins the two given points corresponds to a maximum, for we may suppose that p is positive.

When $\phi'(p) = 0$ we have $p = \pm 1$; and these two values give opposite signs to $\phi''(p)$, so that we do not obtain either a maximum or a minimum by combining these two straight lines.

But $p = \infty$ is also a solution of $\phi'(p) = 0$; and it will be found that by combining $p = 1$ with $p = \infty$ we obtain a maximum, and in fact the greatest value of the proposed integral.

No minimum presents itself in this case. In fact we can make $\int \phi(p)\,dx$ as small as we please. For taking the diagram of the preceding Article we have $\phi(p)$ vanishing along AC and CB: and we can draw a curve very near to AC and CB for which $\int \phi(p)\,dx$ will be as small as we please. But we do not obtain a *minimum* in the technical sense of the word; the integral can be made to change sign as it passes through zero.

9. III. Suppose $\phi(p) = b^2 - \frac{a^2p^2}{2} + \frac{p^4}{4}$.

Then $\phi'(p) = p(p^2 - a^2)$, $\phi''(p) = 3p^2 - a^2$.

Here the straight line which joins the two given points corresponds to a maximum or a minimum according as the value of p^2 is less or greater than $\frac{a^2}{3}$.

The solution made up of the two straight lines AC and BC which are determined by $p = \pm a$ corresponds to a minimum.

Or we may form a solution by combining more than two straight lines, if for every one of them $p^2 = a^2$, so that $p = \pm a$. The value of the integral is the same whatever be the number of *tacks* comprised in the solution.

In continuing the discussion we will for simplicity put $a = 1$.

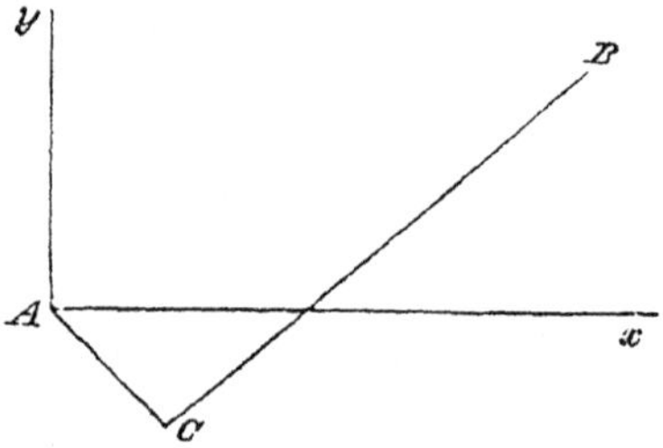

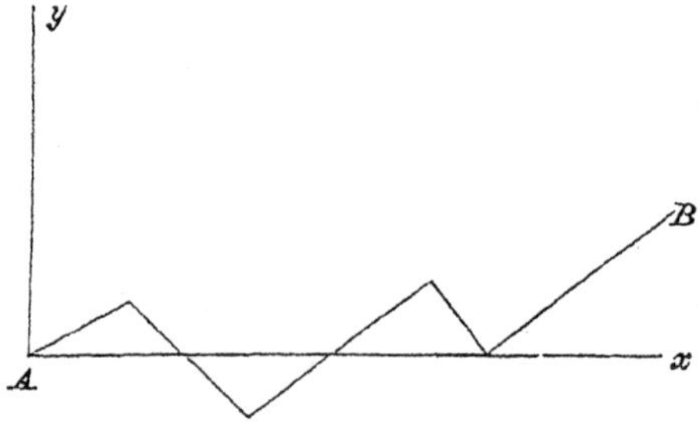

Then in this example if the angle between BA and Ax is greater than 30°, we get a minimum either by the straight line from A to B or by a *tack*. We may remark in passing that it is rare to obtain two minima solutions of a problem in the Calculus of Variations, or rather has been hitherto rare: we shall see other examples. To determine which of these two corresponds to the less result we must determine whether $b^2 - \frac{p^2}{2} + \frac{p^4}{4}$ is greater or less than $b^2 - \frac{1}{2} + \frac{1}{4}$, that is whether $p^4 - 2p^2 + 1$ is greater or less than 0; it is obvious that $(p^2-1)^2$ is greater than 0: hence the value of the integral is less for the solution with the *tack* than for the solution which consists of one straight line.

We may put $\phi(p)$ in this form

$$\phi(p) = b^2 - \frac{1}{4} + \frac{1}{4}(p^2-1)^2;$$

and then it is obvious that the *least* value of $\int \phi(p)\, dx$ is always obtained by supposing $p^2 = 1$.

If the angle between BA and Ax is less than 30^0 there is only one minimum solution, namely that with the *tack*.

No maximum presents itself except in the case in which the angle between BA and Ax is less than 30^0.

10. In the preceding Article the straight lines which form the *tacks* are *equally* inclined to the axis of x: but this need not necessarily be the case in other examples. See Art. 8.

If the equation $\phi'(p) = 0$ furnishes us with roots numerically unequal we may have straight lines forming tacks which are not equally inclined to the axis of x. This plurality of solutions is well illustrated hereafter in the solid of minimum resistance with a given surface.

If $\phi(p)$ is always positive and finite and cannot vanish, $\int \phi(p)\, dx$ must be susceptible of a minimum value. Suppose that $\phi'(m) = 0$ and that $\phi''(m)$ is *negative*, then $p = m$ does not give us a minimum. In this case the equation $\phi'(p) = 0$ must have two other roots besides m, one greater and one less than m. This is easily illustrated geometrically, supposing p the abscissa and $\phi(p)$ the corresponding ordinate of a curve.

The point M is that which corresponds to $\phi'(m) = 0$ and $\phi''(m)$ negative. It may happen, as in the lower diagram, that $p = \pm \infty$ for the other roots of $\phi'(p) = 0$.

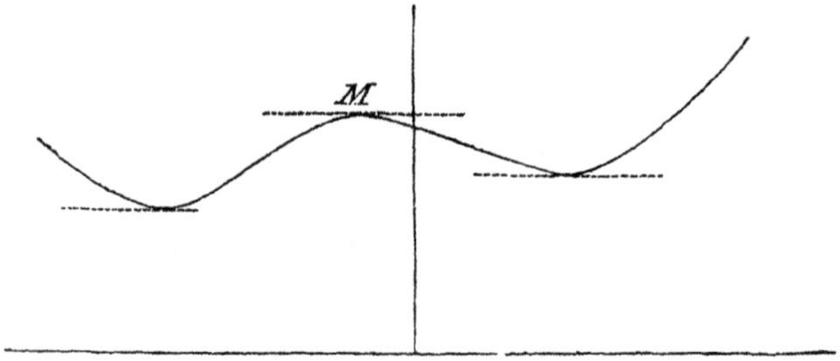

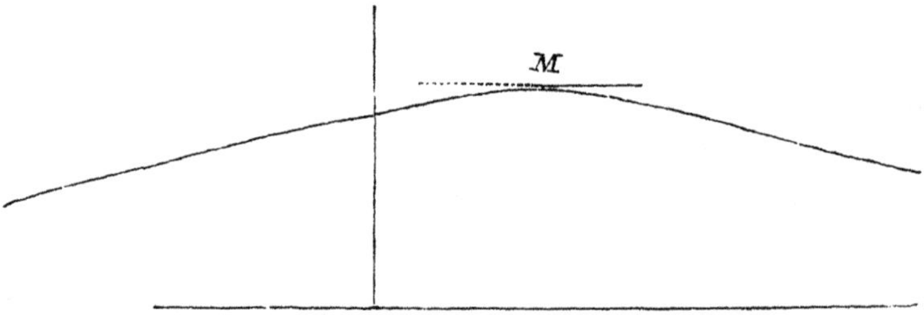

11. Thus we see that the discontinuity which occurs in some cases of the problem of making $\int \phi(p)\, dx$ a maximum or a minimum is that of two or more straight lines meeting at an angle.

12. A particular case of this problem has been considered in a paper on the *Brachistochronous Course of a Ship*: see *Philosophical Magazine* for January 1834, pages 33...36. This paper is I believe the first in which the interesting kind of discontinuity we are here considering was noticed; other kinds of discontinuity had already been discussed, as for instance some by Legendre.

We may suppose the axis of x to coincide with the direction of the wind. The velocity of a ship may be supposed to be a function of the tangent of the angle which the direction of the ship's course makes with the direction of the wind; and from the nature of the case this function must be an *even* function, so that we may denote it by $f(p^2)$. Thus we require the minimum value of the integral $\int \frac{\sqrt{(1+p^2)}}{f(p^2)} dx$, the limits being supposed fixed.

Put $\phi(p)$ for $\frac{\sqrt{(1+p^2)}}{f(p^2)}$; then we get

$$\phi'(p) = p\frac{f - 2(1+p^2)f'}{f^2(1+p^2)^{\frac{1}{2}}}$$

$$\phi''(p) = \frac{f^2 - 2(1+p^2)(1+3p^2)ff' + 8p^2(1+p^2)^2(f')^2 - 4p^2(1+p^2)^2ff''}{f^3(1+p^2)^{\frac{3}{2}}}.$$

where f, f', and f'' are used for brevity instead of $f(p^2)$, $f'(p^2)$, and $f''(p^2)$.

13. We may observe that $f(p^2)$ will in general involve radicals with ambiguities of signs; for the velocity will not be the same for two angles between the ship's course and the direction of the wind which are *supplemental*, although p^2 will have the same value for the two angles. Suppose, for example, that the velocity varies as the square of the cosine of half the angle between the ship's course and the direction of the wind; then $f(p^2)$ varies as $1 \pm \frac{1}{\sqrt{(1+p^2)}}$ where the upper or the lower sign must be taken according as the angle between the ship's course and the direction towards which the wind blows is less or greater than a right angle.

We may observe also that $\sqrt{(1+p^2)}$ must be taken *negatively* in the integral $\int \frac{\sqrt{(1+p^2)}}{f(p^2)} dx$ whenever x is algebraically decreasing.

In order that the value $p = 0$ may correspond to a minimum we must have $f(0) - 2f'(0)$ positive.

14. I shall now make some remarks on the problem which has hitherto been discussed; the second and the third remarks I consider of peculiar importance, for as we proceed it will be found that they are applicable to many other problems.

I. It may be said that in a certain sense there is *no discontinuity;* for example in Art. 9 we obtained two straight lines meeting at an angle; now two straight lines might be regarded as a *conic section*, that is as forming but *one* curve. I lay no stress on this remark, merely introducing it to shew that it has not been overlooked. In fact the solution resembles the double solution of a quadratic, from which indeed it arises.

II. Such discontinuity as occurs may be said to be introduced by the conditions which we impose on the problem. We require a line to have a certain minimum property and to connect *two* fixed points; and the discontinuity arises from the circumstance of there being *two* fixed points. Suppose we take the following problem: find a line such that $\int \phi(p)\,dx$ may be a maximum or a minimum, the line commencing at a given point and having a given length. Here by the usual theory we have to find a maximum or minimum value of $\int \left\{\phi(p) + \lambda\sqrt{(1+p^2)}\right\} dx$, where λ is some constant to be determined. Thus we obtain

$$\phi'(p) + \frac{\lambda p}{\sqrt{(1+p^2)}} = \text{constant};$$

and in order that the term in the variation which is outside the integral sign may vanish this constant must be zero. Also $\phi(p) + \lambda\sqrt{(1+p^2)}$ must be zero at the limit of the integration which is not fixed. Thus we can eliminate λ and we have for determining p the equation

$$\phi'(p) = \frac{p\phi(p)}{1+p^2}.$$

Then the unfixed limit of integration will be determined from the fact that the length of the line is given. Here no discontinuity presents itself. The last equation may furnish us with more than one value of p, but we shall not be able to combine two different values into one solution, unless indeed the two values of p which we employ give the same value to $\frac{\phi(p)}{\sqrt{1+p^2}}$, and also give the same value to $\frac{\phi'(p)\sqrt{1+p^2}}{p}$.

III. A question may naturally occur as to the possibility of finding a solution of our original problem which does not involve discontinuity. Granting that in a particular case the *least* value of the integral is obtained by taking the locus consisting of the

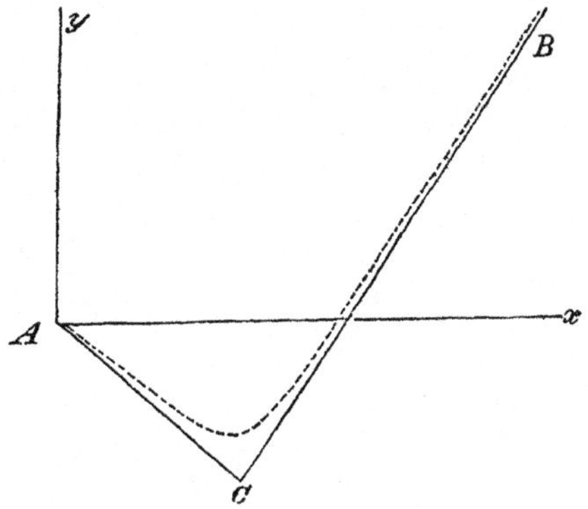

straight lines AC and CB, it might be asked what *curve* proceeding from A to B without any abrupt change of direction will give the least value among all such *curves*. I say in reply, most decidedly, that it is hopeless to seek for such a curve. For we may draw a curve as close as we please to the locus formed of the two straight lines, and thus obtain a result as near as we please to the absolutely least result. The dotted line in the diagram is intended to represent a curve drawn very close to the straight lines.

15. I now proceed to another problem.

Let q stand for $\frac{d^2y}{dx^2}$; and let $\phi(q)$ denote a given function of q: required the curve for which the integral $\int \phi(q)\,dx$ taken between fixed limits is a maximum or a minimum.

Let $u = \int \phi(q)\,dx$; then, as far as terms of the second order inclusive,

$$\delta u = \int \left\{\phi'(q)\,\delta q + \phi''(q)\,\frac{(\delta q)^2}{2}\right\} dx$$

$$= \phi'(q)\,\delta p - \left\{\frac{d}{dx}\phi'(q)\right\}\delta y + \int \left\{\delta y \frac{d^2}{dx^2}\phi'(q) + \phi''(q)\frac{(\delta q)^2}{2}\right\} dx.$$

Then we require by the usual theory

$$\frac{d^2}{dx^2}\phi'(q) = 0;$$

therefore $$\phi'(q) = Cx + C',$$
where C and C' are constants.

Then in order to make the term $\phi'(q)\,\delta p$ which is outside the integral sign vanish it is obvious that we must have $C=0$ and $C'=0$; supposing the lower limit of x to be zero. Hence
$$\phi'(q) = 0;$$
thus one or more constant values can be obtained for q. Suppose a to denote one of these values; then $\frac{d^2y}{dx^2} = a$; therefore
$$y = \frac{ax^2}{2} + bx + b',$$
where b and b' are constants. These constants can be determined from the fact that the curve is to pass through two fixed points. Thus we obtain a parabola for the required curve. And δu is thus reduced to $\frac{\phi''(a)}{2}\int(\delta q)^2\,du$, so that we have a minimum or a maximum according as $\phi''(a)$ is positive or negative.

In this problem then there is no discontinuity; we find on examination that we can satisfy all the conditions by *one* parabola. But we may easily modify the problem so as to introduce discontinuity. For example, suppose we require that the curve shall not only have fixed extremities, say A and B, but also pass through another given point D. If D happens to be on the parabola which passes through A and B and has the maximum or minimum property we have no discontinuity in our solution; but if D be not on this parabola the required curve will consist of two parabolic arcs, one passing from A to D, and the other from D to B; or it may be one passing from A to B and the other from B to D.

16. Thus we see by examples that discontinuity may be produced by conditions imposed on the problems; sometimes unconsciously imposed as in the problem of Art. 2, and sometimes consciously imposed as in the latter part of Art. 15. And as we shall see hereafter conditions may present themselves very naturally which produce great discontinuity in solutions.

But before we discuss other problems it will be convenient to give some general theoretical investigations.

CHAPTER II.

THEORETICAL INVESTIGATIONS.

17. LET there be an integral $\int \phi \, dx$ which is required to be a maximum or a minimum, where ϕ is a known function of y and its differential coefficients with respect to x. Change y into $y + \delta y$; then in the usual way we obtain for the variation of the integral to the first order an expression of the form

$$L + \int M \delta y \, dx,$$

where L depends on the values of the variables and the differential coefficients at the limits of the integration. Now if δy may have either sign we must have $M = 0$ as an indispensable condition for a maximum or a minimum; and moreover we must also have $L = 0$. These statements are universally admitted to be true.

Suppose however that owing to some condition in the problem we cannot always give to δy either sign: for example suppose that throughout the whole range of the integration δy is *essentially positive,* then it is no longer necessary that M should vanish. If M is positive through the whole range of the integration we are sure of a minimum; and if M is negative through the whole range of the integration we are sure of a maximum. We assume of course that we are able to satisfy the condition $L = 0$; or to ensure that L shall be positive in the former case and negative in the latter case.

Next suppose that δy may have either sign through part of the range of the integration, but that it is essentially positive during the remainder of the range. Then if M vanishes through the former part and is positive through the latter part of the range we are sure of a minimum; and if M vanishes through the former part and is negative through the latter part of the range we are sure of a maximum. We assume as before that the condition relating to L can be satisfied.

Now we must observe a great peculiarity in the case which we are considering; when δy does not vanish throughout the range for which its sign is restricted we are sure that the variation of the integral is essentially positive or essentially negative without examining the terms of the *second order* in the variation.

18. Simple as the remark is which is the subject of the preceding Article it will furnish the foundation for much that will follow: in fact it is the principle on which depends the discontinuous solution of nearly all the problems we shall have to discuss. The principle appears to have been first employed by Mr Todhunter in the *Philosophical Magazine* for June 1866. For the applications we shall have to make of the principle we may state it thus: Suppose we are seeking by the aid of the Calculus of Variations the curve which has some assigned maximum or minimum property; then if no condition is imposed which fetters the sign of δy there can be no solution except such as may be supplied by putting $M=0$. But suppose a certain boundary is imposed which the curve is not to pass beyond; then along that boundary δy will not be susceptible of both signs, so that part or the whole of this boundary may occur in the required solution. Thus the solution can consist of nothing besides what can be obtained from $M=0$, or of part or the whole of the given boundary, or of some combination of these two elements. Many illustrations of this statement will occur as we proceed.

19. The integrals with which we shall be concerned will in general have to be taken between assigned limits. Hence the variation which we have denoted by $L+\int M\,\delta y\,dx$ is more ex-

plicitly presented thus,

$$L_1 - L_0 + \int_{x_0}^{x_1} M\delta y\, dx,$$

where L_1 denotes the value of a certain expression at the upper limit of integration, and L_0 the value of the same expression at the lower limit.

Now suppose we separate our range of integration into two parts, one extending from x_0 to ξ and the other from ξ to x_1. Then the variation corresponding to the range from x_0 to ξ may be denoted by

$$L_2 - L_0 + \int_{x_0}^{\xi} M\delta y\, dx,$$

and the variation corresponding to the range from ξ to x_1 may be denoted by

$$L_1 - L_3 + \int_{\xi}^{x_1} M\delta y\, dx.$$

If there be no discontinuity in the function to which variation has been given L_2 and L_3 will really denote the same thing; but if there is discontinuity L_2 and L_3 will not necessarily denote the same thing: and we shall have to be very careful on this point when we are considering the value or the sign of the whole variation.

20. A simple example may be here conveniently solved, as it will illustrate some of the remarks made in the present Chapter.

Required a curve which shall connect two fixed points A and B on the axis of x, and make $\int (q^2 - 2y)\, dx$ a minimum.

Let $$u = \int (q^2 - 2y)\, dx;$$

then to the second order inclusive we have

$$\begin{aligned} \delta u &= \int (2q\delta q - 2\delta y)\, dx + \int (\delta q)^2\, dx \\ &= 2q\delta p - 2\frac{dq}{dx}\delta y + 2\int \left(\frac{d^2q}{dx^2} - 1\right)\delta y dx \\ &\qquad + \int (\delta q)^2\, dx. \end{aligned}$$

Hence we have in the usual way

$$\frac{d^2q}{dx^2} - 1 = 0 \qquad (1),$$

whence
$$y = \frac{x^4}{24} + Cx^3 + C'x^2 + C''x + C'''.$$

Let $AB = 2h$; and take the middle point between A and B

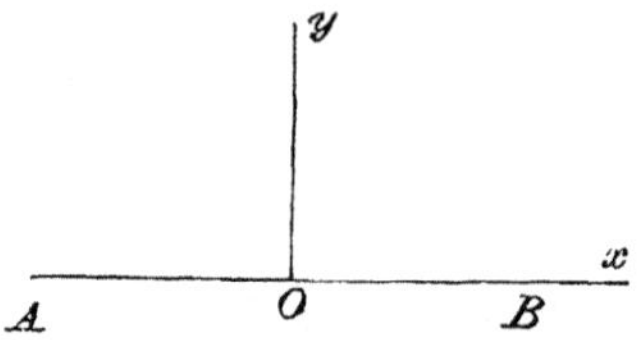

for the origin. Then to make the term $2q\,\delta p$ vanish we must have $q = 0$ both when $x = h$ and when $x = -h$: this leads to $C = 0$ and $C' = -\frac{h^2}{4}$. Also $y = 0$ when $x = h$ and when $x = -h$: this leads to $C'' = 0$ and $C''' = \frac{5h^4}{24}$. Thus finally

$$y = \frac{x^4}{24} - \frac{h^2x^2}{4} + \frac{5h^4}{24} \qquad (2).$$

But suppose we impose the condition that y is to be always positive. Then along the axis of x the sign of δy can never be negative. Thus we are led to enquire whether $y = 0$ is not a solution: we find however that this is not a solution. For the term of the first order in δu now reduces to $-2\int \delta y dx$, and as δy cannot be negative this term is negative; so that u is not a minimum.

If however we impose the condition that y is always to be negative $y = 0$ is a solution.

Similarly if any curve be drawn from A to B, and the condition be imposed that the required curve is not to fall between this given curve and the straight line AB, then if $\frac{d^2q}{dx^2} - 1$ is positive for all points of the given curve this curve itself supplies a minimum solution. If the condition be imposed that the required curve is not to fall beyond the given curve, then if $\frac{d^2q}{dx^2} - 1$ is negative for all points of the given curve this curve itself supplies a minimum solution.

Now let us take the problem as originally proposed with this condition; that a certain given point is not to be excluded by the curve, so that this point is to be either within the given curve or on it.

Let the co-ordinates of the given point be a and b; let D denote the point. The problem of course is by no means the same as if we *required* the curve to pass through D. If we required the curve to pass through D we should in fact have to draw one curve between the fixed points A and D having the assigned minimum property, and another curve between the fixed points D and B having the assigned minimum property: then in fact we should have two problems identical in principle with that which we have already solved. But the problem we have really to solve is different.

If the point D falls within the curve (2) then that curve is the solution required. But if the point D does not fall within the curve (2) no solution can exist except one which has the point D on it: we proceed to seek this solution.

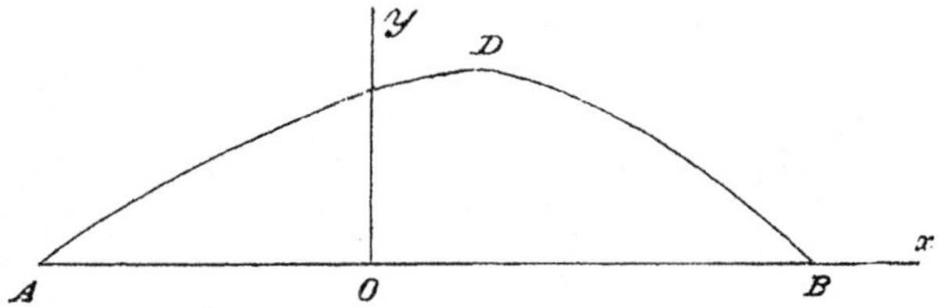

First consider the portion DB.

This must be determined by the differential equation (1). Hence we must have

$$y = \frac{x^4}{24} + Cx^3 + C'x^2 + C''x + C''' \text{ .... (3).}$$

Now as the term $2q\,\delta p$ must vanish at B we must have $q=0$ when $x=h$, so that

$$\frac{h^2}{2}+6Ch+2C'=0 \qquad \text{.......................(4)}.$$

Also the curve is to pass through B and D; thus

$$\left.\begin{aligned} 0&=\frac{h^4}{24}+Ch^3+C'h^2+C''h+C''' \\ b&=\frac{a^4}{24}+Ca^3+C'a^2+C''a+C''' \end{aligned}\right\} \qquad \text{............(5)}.$$

From the equations (4) and (5) we may express C', C'', and C''' in terms of C and known quantities; so that C alone remains undetermined.

In the same manner we find for AD the equation

$$y=\frac{x^4}{24}+\gamma x^3+\gamma' x^2+\gamma'' x+\gamma''';$$

the constants γ, γ', γ'', and γ''' being connected by relations like (4) and (5), with the sign of h changed; thus we may consider that γ alone remains undetermined.

Now let us examine the relations which must hold at the point D. We must have with the notation of Art. 19

$$(2q\,\delta p)_2-(2q\,\delta p)_3=0,$$

where the subscript 2 relates to the point D as being on the arc AD, and the subscript 3 relates to the point D as being on the arc BD. The only way to make this vanish always [unless both q_2 and q_3 are zero] is to have $q_2=q_3$, and also $\delta p_2=\delta p_3$: the latter requires that $p_2=p_3$, for then, and then only, δp_2 and δp_3 mean the same thing. [This however implies that we assume there is to be no break of direction at D.] Hence we must have

$$(6)\left\{\begin{aligned} &\frac{a^2}{2}+6aC+2C'=\frac{a^2}{2}+6a\gamma+2\gamma' \\ &\frac{a^3}{6}+3a^2C+2aC'+C''=\frac{a^3}{6}+3a^2\gamma+2a\gamma'+\gamma''. \end{aligned}\right.$$

These equations with those which have been previously obtained suffice to determine all the arbitrary constants.

We have still left in the variation of the integral the term of the first order

$$\left(-2\frac{dq}{dx}\delta y\right)_2 - \left(-2\frac{dq}{dx}\delta y\right)_3;$$

here δy_2 and δy_3 mean the same thing, so that the term reduces to

$$6(C-\gamma)\,\delta y_2.$$

Now by the nature of the problem δy_2 can never be negative. Hence we are certain of a minimum if $C-\gamma$ is positive, for then this term is positive or zero; and the term of the second order is positive.

From (4)

$$C' = -3Ch - \frac{h^2}{4}.$$

Similarly $$\gamma' = 3\gamma h - \frac{h^2}{4}.$$

Substitute in the first of equations (6); thus

$$C(a-h) = \gamma(a+h),$$

so that $$C-\gamma = \frac{2hC}{a+h};$$

we have then to shew that C is positive.

From (5)

$$C'' + C'(a+h) + C(a^2+ah+h^2) + \frac{a^3+a^2h+ah^2+h^3}{24} = \frac{b}{a-h}.$$

Substitute for C'; thus

$$C'' - (a+h)\left(3Ch + \frac{h^2}{4}\right) + C(a^2+ah+h^2) + \frac{a^3+a^2h+ah^2+h^3}{24} = \frac{b}{a-h},$$

that is

$$C'' + C(a^2-2ah-2h^2) + \frac{a^3+a^2h-5ah^2-5h^3}{24} = \frac{b}{a-h} \quad\ldots\ldots\ldots (7).$$

Similarly

$$\gamma'' + \gamma(a^2+2ah-2h^2) + \frac{a^3-a^2h-5ah^2+5h^3}{24} = \frac{b}{a+h} \quad\ldots (8).$$

From (7) and (8) by subtraction

$$C'' - \gamma'' + C\left\{a^2 - 2ah - 2h^2 - \frac{a-h}{a+h}(a^2 + 2ah - 2h^2)\right\} + \frac{a^2h - 5h^3}{12}$$

$$= \frac{2bh}{a^2 - h^2},$$

that is $$C'' - \gamma'' - \frac{2a^2h + 4h^3}{a+h}C + \frac{a^2h - 5h^3}{12} = \frac{2bh}{a^2 - h^2}.$$

But from the second of equations (6) we get

$$C'' - \gamma'' = 2a(\gamma' - C') + 3a^2(\gamma - C)$$

$$= 6ah(\gamma + C) + 3a^2(\gamma - C)$$

$$= \frac{6a^2h}{a+h}C,$$

so that $$\frac{4(a^2h - h^3)}{a+h}C + \frac{a^2h - 5h^3}{12} = \frac{2bh}{a^2 - h^2};$$

therefore $$4C(h-a) = \frac{2b}{h^2 - a^2} - \frac{5h^2 - a^2}{12}.$$

Thus C is positive provided $2b$ is greater than

$$\frac{(5h^2 - a^2)(h^2 - a^2)}{12},$$

that is provided b is greater than

$$\frac{a^4}{24} - \frac{a^2h^2}{4} + \frac{5h^4}{24};$$

and this condition is satisfied inasmuch as D is supposed to be outside the curve (2).

It will be observed that the solution is discontinuous; the branches which meet at D are determined by different equations; at the common point p and q have respectively the same value for each branch. The result obtained is a *minimum;* it is not the *least* value of the proposed integral.

It is obvious that the integral can be made arithmetically as small as we please by taking parts of the axis of x and two straight lines inclined to the axis of x at angles very nearly right angles: this is illustrated by the parts AG, GD, DH, HB of the diagram.

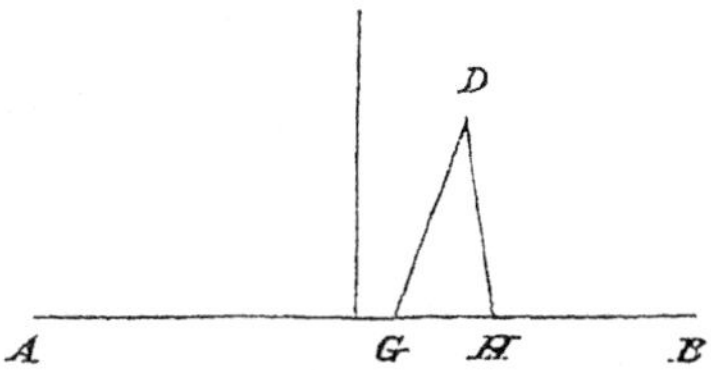

[The result obtained is a minimum subject to the condition introduced in the investigation. We may enunciate the result thus: if any other curve be drawn by giving to y and q infinitesimal changes, so as not to exclude D, and to have no break of direction at D, the integral has an algebraically greater value than it has for the assigned solution.

We may briefly notice another case which presented itself. Suppose we put $q_2 = 0$, and $q_3 = 0$. Then we obtain

$$C = -\frac{a+h}{12}, \quad \gamma = -\frac{a-h}{12};$$

therefore

$$(C-\gamma)\,\delta y_2 = -\frac{h}{6}\,\delta y_2.$$

Hence this case does not give a minimum.]

21. It will be useful for the sake of reference to present some known formulæ with respect to terms of the second order in a variation; I shall add something to the usual investigations of them. I confine myself, as sufficient for my purpose, to the case in which the function under the integral sign involves no differential coefficient higher than the first.

Let then $u = \int \phi(x, y, p)\,dx$, where as usual p stands for $\frac{dy}{dx}$. Change y into $y + \delta y$; then to terms of the second order inclusive we have

$$\delta u = \int \left(\frac{d\phi}{dy}\,\delta y + \frac{d\phi}{dp}\,\delta p \right) dx$$

$$+ \frac{1}{2} \int \left\{ \frac{d^2\phi}{dy^2} (\delta y)^2 + 2\,\frac{d^2\phi}{dy\,dp}\,\delta y\,\delta p + \frac{d^2\phi}{dp^2} (\delta p)^2 \right\} dx.$$

Denote the term of the first order in δu by v, and the term of the second order by $\frac{1}{2}w$; then we see that

$$w = \delta v:$$

and we may conclude that this relation will still hold after v has been transformed in any convenient manner.

Now if we transform v in the usual manner, using the subscripts 1 and 0 to denote values at the upper and the lower limits of integration, we obtain

$$v = \left(\frac{d\phi}{dp}\delta y\right)_1 - \left(\frac{d\phi}{dp}\delta y\right)_0 + \int \left\{\frac{d\phi}{dy} - \frac{d}{dx}\left(\frac{d\phi}{dp}\right)\right\} \delta y\, dx.$$

We conclude then that w is the variation of this expression; that is we conclude that

$$w = \left\{\frac{d^2\phi}{dy\,dp}(\delta y)^2 + \frac{d^2\phi}{dp^2}\delta y\,\delta p\right\}_1 - \left\{\frac{d^2\phi}{dy\,dp}(\delta y)^2 + \frac{d^2\phi}{dp^2}\delta y\,\delta p\right\}_0$$

$$+ \int \left\{\frac{d^2\phi}{dy^2}\delta y + \frac{d^2\phi}{dy\,dp}\delta p - \frac{d}{dx}\left(\frac{d^2\phi}{dy\,dp}\delta y + \frac{d^2\phi}{dp^2}\delta p\right)\right\} \delta y\, dx;$$

of which the last line may be written

$$\int \left\{\left[\frac{d^2\phi}{dy^2} - \frac{d}{dx}\left(\frac{d^2\phi}{dy\,dp}\right)\right]\delta y - \frac{d}{dx}\left(\frac{d^2\phi}{dp^2}\delta p\right)\right\} \delta y\, dx.$$

This is the form in which it is found convenient to put the term of the second order in the variation according to the known method of Jacobi. It is easy to obtain this form directly instead of indirectly as we have done; as we will now shew.

22. We have

$$w = \int \left\{\frac{d^2\phi}{dy^2}(\delta y)^2 + 2\frac{d^2\phi}{dy\,dp}\delta y\,\delta p + \frac{d^2\phi}{dp^2}(\delta p)^2\right\} dx.$$

By integration by parts we have

$$\int \frac{d^2\phi}{dp^2}(\delta p)^2\, dx = \frac{d^2\phi}{dp^2}\delta y\,\delta p - \int \delta y \frac{d}{dx}\left(\frac{d^2\phi}{dp^2}\delta p\right) dx.$$

Also by integration by parts we have

$$\int \frac{d^2\phi}{dy\,dp}\,\delta y\,\delta p\,dx = \frac{d^2\phi}{dy\,dp}(\delta y)^2 - \int \delta y\,\frac{d}{dx}\left(\frac{d^2\phi}{dy\,dp}\,\delta y\right)dx$$

$$= \frac{d^2\phi}{dy\,dp}(\delta y)^2 - \int \delta y\,\delta p\,\frac{d^2\phi}{dy\,dp}\,dx - \int (\delta y)^2\,\frac{d}{dx}\left(\frac{d^2\phi}{dy\,dp}\right)dx\,;$$

therefore $2\int \frac{d^2\phi}{dy\,dp}\,\delta y\,\delta p\,dx = \frac{d^2\phi}{dy\,dp}(\delta y)^2 - \int (\delta y)^2\,\frac{d}{dx}\left(\frac{d^2\phi}{dy\,dp}\right)dx.$

Thus we obtain the required transformation of w.

23. We shall now suppose that the limiting of values of y are fixed, so that $\delta y_1 = 0$ and $\delta y_0 = 0$.

Hence we have

$$w = \int \left\{P\,\delta y - \frac{d}{dx}(Q\,\delta p)\right\}\delta y\,\delta x,$$

where $\qquad P$ stands for $\frac{d^2\phi}{dy^2} - \frac{d}{dx}\left(\frac{d^2\phi}{dy\,dp}\right)$

and $\qquad Q$ stands for $\frac{d^2\phi}{dp^2}$.

Let z be such a quantity that

$$Pz - \frac{d}{dx}\left(Q\,\frac{dz}{dx}\right) = 0 \;\ldots\ldots\ldots\ldots\ldots\ldots\; (1),$$

and assume $\delta y = tz$; then

$$\delta y\left\{P\,\delta y - \frac{d}{dx}(Q\delta p)\right\}$$

$$= \delta y\left\{Ptz - \frac{d}{dx}\left(Q\,\frac{d}{dx}[tz]\right)\right\}$$

$$= tz\left\{t\,\frac{d}{dx}\left(Q\,\frac{dz}{dx}\right) - \frac{d}{dx}\,Q\,\frac{d}{dx}[tz]\right\}$$

$$= -t\,\frac{d}{dx}\left(Qz^2\,\frac{dt}{dx}\right):$$

therefore $\int \delta y\left\{P\,\delta y - \frac{d}{dx}(Q\,\delta p)\right\}dx = -\int t\,\frac{d}{dx}\left(Qz^2\,\frac{dt}{dx}\right)dx$

$$= -\,tQz^2\,\frac{dt}{dx} + \int Qz^2\left(\frac{dt}{dx}\right)^2 dx.$$

Since the limiting values of y are fixed the part of the last expression which is outside the integral sign vanishes. Hence finally the term of the second order in δu

$$=\frac{1}{2}\int Qz^2\left(\frac{dt}{dx}\right)^2 dx=\frac{1}{2}\int Q\left(\frac{z\delta p-z'\delta y}{z}\right)^2 dx.$$

Here z' is used for $\frac{dz}{dx}$; and in like manner y' will be used for $\frac{dy}{dx}$, and y'' for $\frac{d^2y}{dx^2}$.

This is Jacobi's form for the result. We see at once that there will be neither a maximum nor a minimum unless Q preserves the same sign throughout the range of the integration.

24. Suppose then that Q does retain the same sign throughout the range of the integration. We shall now consider what further conditions are necessary in order to ensure a maximum or a minimum: this is a point which the ordinary treatises on the subject seem to me to discuss in an unsatisfactory manner.

I. Suppose that z can be taken so as never to vanish throughout the range of the integration; then there is a maximum if Q is negative and a minimum if Q is positive. For we see that the expression under the integral sign in the term of the second order in δu is necessarily of the same sign as Q; and it will not vanish unless throughout the range of integration $z\delta p-z'\delta y=0$, which we may write thus $z\delta y'-z'\delta y=0$. But this leads to $\delta y=Cz$, where C is a constant, and as δy vanishes at the limits z must also vanish, which is contrary to the supposition.

Now as we shall see presently z is of the form $C_1f_1+C_2f_2$ where C_1 and C_2 are arbitrary constants, and f_1 and f_2 are definite functions of x. If f_1 and f_2 do not vanish nor become infinite within the range of integration we can secure that z shall not vanish. For

$$z=C_1(f_1+mf_2),$$

where $m=\frac{C_2}{C_1}$; and $\frac{f_1}{f_2}$ will not range from positive infinity to negative infinity, but only between certain finite limits; so that by ascribing to m any value outside these limits we secure that z shall not vanish.

II. Suppose that we cannot secure that z shall not vanish throughout the range of integration. Our assumption that $\delta y = tz$ is not admissible if z vanishes when δy does not vanish. Hence it might appear at first sight that the proposed method of transformation simply becomes inapplicable, and so leads to no result. But as we shall shew we can infer that there is now neither a maximum nor a minimum. For from what has previously been said it follows that $\frac{f_1}{f_2}$ will now range through every value from positive infinity to negative infinity. Take m such that z shall vanish at the lower limit of integration. Then by reason of the range of values of which $\frac{f_1}{f_2}$ is susceptible z will also vanish at or before the upper limit of integration. Take $\delta y = Cz$, where C is a constant, for all values of the variable x between those for which z vanishes: and take $\delta y = 0$ for other values of x. Then the term of the second order in δu vanishes. The term of the third order will in general not vanish, but will be susceptible of either sign. Thus there is neither a maximum nor a minimum.

25. Suppose that the value of y found from

$$\frac{d\phi}{dy} - \frac{d}{dx}\left(\frac{d\phi}{dp}\right) = 0 \qquad\qquad (2)$$

is denoted by $f(x, c_1, c_2)$ where c_1 and c_2 are arbitrary constants: then this value of y is of course the solution of the problem of the Calculus of Variations which is supposed to be under discussion.

Equation (2) will also be satisfied when we give small arbitrary increments δc_1 and δc_2 to the constants. And as we shewed in Art. 21 that $w = \delta v$ we infer that (1) will be satisfied when we put

$$z = \frac{df}{dc_1}\,\delta c_1 + \frac{df}{dc_2}\,\delta c_2.$$

And as (1) is linear with respect to z we see finally that the general value of z is

$$z = C_1 \frac{df}{dc_1} + C_2 \frac{df}{dc_2},$$

where C_1 and C_2 are arbitrary constants.

This finishes the exposition of Jacobi's method for discriminating between a maximum and a minimum, so far as will be necessary for our purpose.

26. I propose however to consider more particularly the term of the second order in the variation of $\int y\,\phi(p)\,dx$. This is of course less general than the problem which has just been given after Jacobi; but it includes a large number of particular cases, and it will furnish some results which have not hitherto been specially noticed.

Let $$u = \int y\,\phi(p)\,dx\,;$$

then to terms of the second order inclusive we have

$$\delta u = \int \{\delta y\,\phi(p) + y\delta p\,\phi'(p)\}\,dx$$
$$+\frac{1}{2}\int \{2\delta y\,\delta p\,\phi'(p) + (\delta p)^2\,y\,\phi''(p)\}\,dx.$$

Transform the term of the first order in the ordinary way, and suppose the limiting values of y to be fixed; then this term becomes

$$\int \left\{\phi(p) - \frac{d}{dx}[y\,\phi'(p)]\right\}\delta y\,dx.$$

To make this vanish we must have

$$\phi(p) - \frac{d}{dx}[y\,\phi'(p)] = 0,$$

and this leads to

$$y\,\phi(p) - y\,p\,\phi'(p) = c_1 \quad\ldots\ldots\ldots\ldots\ldots\ldots\ldots (3),$$

where c_1 is a constant.

Now consider the term of the second order.

By integrating by parts we have

$$\int \delta y\,\delta p\,\phi'(p)\,dx = (\delta y)^2\,\phi'(p) - \int \delta y\,\frac{d}{dx}[\phi'(p)\,\delta y]\,dx\,;$$

therefore

$$2\int \delta y\,\delta p\,\phi'(p)\,dx = (\delta y)^2\,\phi'(p) - \int (\delta y)^2\,\frac{d}{dx}\,\phi'(p)\,dx\,;$$

and thus the term of the second order is

$$\frac{1}{2}\int\{y\,\phi''(p)\,(\delta p)^2 - y''\,\phi''(p)\,(\delta y)^2\}\,dx,$$

that is

$$\frac{1}{2}\int\phi''(p)\,\{y\,(\delta p)^2 - y''\,(\delta y)^2\}\,dx.$$

This supposes that δy vanishes at the limits; if it does not the term of the second order is

$$\frac{1}{2}\{\phi'(p)\,(\delta y)^2\}_1 - \frac{1}{2}\{\phi'(p)\,(\delta y)^2\}_0$$

$$+\frac{1}{2}\int\phi''(p)\,\{y\,(\delta p)^2 - y''\,(\delta y)^2\}\,dx,$$

where the subscripts 1 and 0 refer to the upper and lower limits of integration respectively.

At present we will however continue to suppose that δy vanishes at the limits.

27. In geometrical applications we shall generally be able to regard y as positive. If then the curve given by (3) is *concave* to the axis of x we know that y'' is negative; and thus the sign of the term of the second order will be invariable if that of $\phi''(p)$ is so. But we can shew that the sign of $\phi''(p)$ is invariable if that of y'' is; for from (3) we have

$$y = \frac{c_1}{\phi(p) - p\phi'(p)},$$

therefore by differentiation

$$p = \frac{c_1\,p\,y''\,\phi''(p)}{\{\phi(p) - p\phi'(p)\}^2},$$

so that

$$y'' = \frac{\{\phi(p) - p\,\phi'(p)\}^2}{c_1\,\phi''(p)}:$$

thus we see that if one of the two y'' and $\phi''(p)$ is of invariable sign so also is the other.

Thus if y be positive and the curve be *concave* to the axis of x we have a maximum if $\phi''(p)$ is negative and a minimum if $\phi''(p)$ is positive: and, in this case we need not have recourse to Jacobi's method.

28. If however the curve be *convex* to the axis of x we cannot settle the sign of the term of the second order without further examination: to this we now proceed.

From (3) we have

$$y = c_1 \psi(p) \quad \ldots\ldots\ldots\ldots (4),$$

where $\psi(p)$ denotes a known function of p.

Now $$x = \int \frac{dx}{dy} dy = \int \frac{dy}{p} = c_1 \int \frac{\psi'(p)}{p} dp.$$

Thus $$x = c_1 \chi(p) + c_2 \quad \ldots\ldots\ldots\ldots (5),$$

where $\chi(p)$ is some definite function of p, and c_2 is an arbitrary constant.

The value of y in terms of x is theoretically to be found by eliminating p between (4) and (5). We denote the result of this elimination by

$$y = f(x, c_1, c_2).$$

Although we cannot actually effect the elimination generally yet we shall be able to obtain the forms of $\frac{df}{dc_1}$ and $\frac{df}{dc_2}$ which are required in Jacobi's method.

For from (4) and (5) we obtain

$$\frac{dy}{dc_1} = \psi(p) + c_1 \psi'(p) \frac{dp}{dc_1},$$

$$0 = \chi(p) + c_1 \chi'(p) \frac{dp}{dc_1}.$$

The second of these equations gives

$$\chi(p) = -c_1 \frac{\psi'(p)}{p} \frac{dp}{dc_1},$$

so that $$\frac{dy}{dc_1} = \psi(p) - p\chi(p) = \frac{y}{c_1} - \frac{p(x - c_2)}{c_1}.$$

Again, from (4) and (5) we have

$$\frac{dy}{dc_2} = c_1 \psi'(p) \frac{dp}{dc_2},$$

$$0 = c_1 \chi'(p) \frac{dp}{dc_2} + 1.$$

The second of these equations gives

$$-1 = \frac{c_1 \psi'(p)}{p} \frac{dp}{dc_2},$$

so that

$$\frac{dy}{dc_2} = -p.$$

Hence the quantity which we denoted by z in Arts. 23 and 25 becomes in the present case

$$C_1 \left\{ \frac{y}{c_1} - \frac{p(x - c_2)}{c_1} \right\} - C_2 p,$$

that is

$$C_1 \left\{ \frac{y}{c_1} - \frac{p(x - c_2)}{c_1} - mp) \right\},$$

where m is a constant standing for $\frac{C_2}{C_1}$.

If the expression just given between brackets vanishes at any point we have at that point

$$x - \frac{y}{p} = c_2 - mc_1.$$

Now $x - \frac{y}{p}$ is the abscissa of the point of intersection of the tangent to the curve at the point (x, y) and the axis of x; we will put ξ for this abscissa.

We assume that y is positive and that the curve is convex to the axis of x.

29. I. Suppose that the tangents at the extreme points of the curve, that is the fixed points, intersect *above* the axis of x. Then ξ is not susceptible of *all* possible values; for instance, if the extreme points of the curve are on opposite sides of the lowest point ξ is not susceptible of values lying between the values it has for the extreme points. Thus we can take $c_2 - mc_1$ so that it shall not be equal to any admissible value of ξ: so that we can secure that z shall not vanish throughout the range of integration. Hence by Art. 24 we are assured of the existence of a maximum or of a minimum if $\phi''(p)$ retains an invariable sign.

II. Suppose that the tangents at the extreme points of the curve do not intersect above the axis of x; then there will be neither a maximum nor a minimum. For if the tangents intersect *on* the axis of x we can make z vanish at the two limits of the integration. If the tangents intersect *below* the axis of x we can make z vanish at one of the limits of integration, and also at some other point within the range of integration. Hence by Art. 24 there is neither a maximum nor a minimum.

Particular cases of this general result have been noticed before; but not the general result itself. See Dienger's *Grundriss der Variationsrechnung*, 1867, pages 21 and 25.

30. It is important to observe that the transformation given by Jacobi for the term of the second order in a variation holds even if we do not suppose the term of the first order to vanish. For instance: let

$$u = \int \phi\,(x, y, p)\,dx\,;$$

then taking the limiting values of the variables to be fixed we obtain

$$\delta u = \int M\delta y\,dx + \frac{1}{2}\int Q\left(\delta p - \frac{z'}{z}\,\delta y\right)^2 dx \;\ldots\ldots\ldots\ldots(1).$$

Now if we put $M = 0$, we have for every system of values of δy

$$\delta u = \frac{1}{2}\int Q\left(\delta p - \frac{z'}{z}\,\delta y\right)^2 dx \;\ldots\ldots\ldots\ldots\ldots\ldots(2).$$

But even if we do not put $M = 0$ the above general value (1) of δu still holds; and in this case it may be possible that some particular value of δy makes $\int M\delta y\,dx = 0$, and then δu reduces as before to the form given in (2).

For example take the case of a brachistochrone under the action of gravity. Here x being measured vertically downwards

$$u = \int \frac{\sqrt{(1+p^2)}}{\sqrt{x}}\,dx.$$

Hence y has to be found from

$$\frac{d}{dx}\frac{p}{\sqrt{x(1+p^2)}}=0.$$

This leads to

$$y=c_1-\sqrt{2c_2x-x^2}+c_2\operatorname{vers}^{-1}\frac{x}{c_2}.$$

Hence

$$z=C_1+C_2\left\{\operatorname{vers}^{-1}\frac{x}{c_2}-\frac{2x}{\sqrt{2c_2x-x^2}}\right\}.$$

Hence with this value of z the expression (1) holds for δu whatever be the relation between x and y. Suppose for illustration that we take a curve consisting of an arc of a circle followed by an arc of a cycloid which has its cusps in the axis of y, the two arcs touching at the common point. Then δu takes the form given in (1). The part of $\int M\delta y\,dx$ which corresponds to the cycloid vanishes; the other part of $\int M\delta y\,dx$ does not vanish always, but it may vanish for some particular value of δy. The term of the second order in δu retains the same form throughout.

Of course it is possible to give special transformations of the term of the second order in a variation in special cases. Thus the transformation in Todhunter's *Integral Calculus*, third edition, Art. 377, applies to the problem there discussed, that is the brachistochrone.

CHAPTER III.

DISCONTINUITY PRODUCED BY CONDITIONS.

31. We shall now discuss some examples in which discontinuity is produced by conditions explicitly imposed on the problems.

We begin with a very simple case.

Imagine a rectangular court containing four rectangular grass

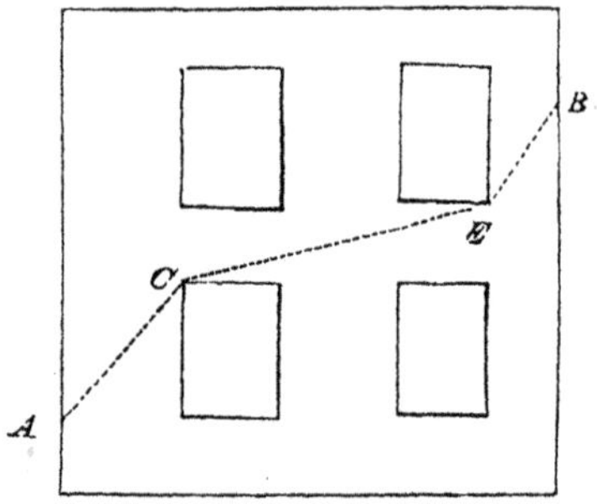

plots. Required the shortest course from a given point A on one side of the court to a given point B on the other side, with the condition that the path is not to cross the grass plots.

Of course A and B might be so situated that a straight line could be drawn from A to B without crossing a grass plot. But if A and B are not so situated the path will be discontinuous, consisting of two or more straight lines not in the same direction.

For instance, in the diagram the shortest path may consist of three parts, namely AC, CE, and EB. In this example the Calculus of Variations will assure us that the path must be made up of straight lines; and then we must determine by Geometry or the Differential Calculus what assemblage of straight lines will constitute the shortest path.

32. We shall now modify the general problem of Art. 2, so as to introduce discontinuity; and as an easy mode of enunciating the problem we will suppose as in Art. 12 that we are dealing with the brachistochronous course of a ship. We will now make the very natural condition that the ship's course is not to cross certain prescribed spaces; these spaces we may conceive to be forbidden on account of rocks, or shoals, or hostile batteries.

Thus, for instance, suppose that a ship is to pass from A to B in the shortest time without crossing the straight line OC pro-

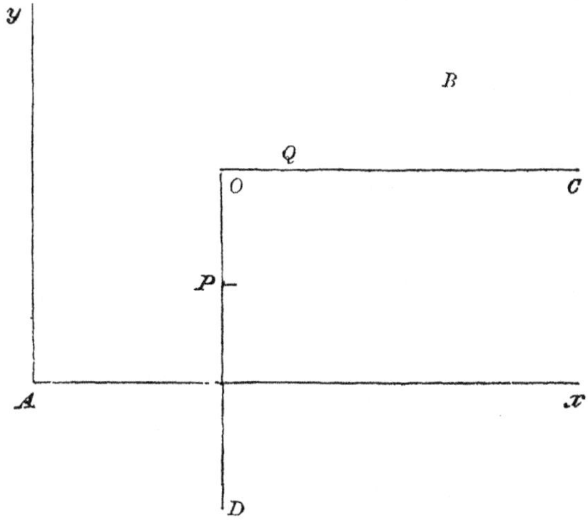

duced indefinitely to the right, or the straight line OD produced indefinitely downwards; these straight lines being parallel to the axes, and Ax being the direction of the wind.

We know from the discussion in Art. 12 that the course between any two points if no obstacle occurs is a straight line, or is composed of more than one straight line; and thus we see that in the present case the course must be composed of various straight lines, including possibly portions of the boundary of the forbidden space. We must then examine the various suppositions that can be made. For example, the swiftest course may perhaps consist of the straight line from A to some point P in OD, the straight line from P to O, the straight line from O to some point Q in OC, and the straight line QB. We should have of course to investigate the positions of P and Q, if such there be, which would make this course swifter than any other.

33. A general solution of the problem of the preceding Article for any form of the function denoted by $\phi(p)$ in Art. 12 would be impracticable. But if a particular form be assigned to ϕ, it might be practicable to complete the discussion: of course the final result might depend on the situation of the point B and of the straight lines OC and OD.

34. We will however discuss a particular case with some detail.

Let the velocity be the function of p which is denoted by $\frac{1}{1+p^2}$, so that $\phi(p)$ in Art. 12 stands for $(1+p^2)^{\frac{3}{2}}$.

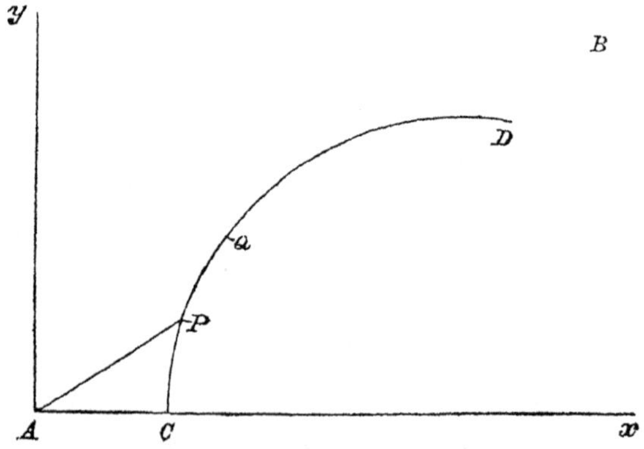

Suppose Ax the direction of the wind; let CD be an arc of a circle: and let the swiftest course be required from A to B with the condition that the course does not cross the circle, which is supposed large enough to forbid the direct course from A to B. The restriction here imposed might present itself in practice owing to the presence of a hostile battery of a certain range.

We have to find the minimum value of

$$\int \phi(p)\,dx \text{ where } \phi(p) = (1+p^2)^{\frac{3}{2}}.$$

Here $$\phi'(p) = 3p\,(1+p^2)^{\frac{1}{2}}.$$

Thus $\phi'(p) = 0$ has no solution except $p = 0$; [and $\phi'(p) = C$ will furnish only one value of p]: and so the swiftest course between any two points is the straight line which joins them; see Arts. 2 and 4. Hence it will follow that if P and Q be adjacent points on the arc the direct course from A to Q is swifter than that made up of the straight line AP and the arc PQ.

We shall by this consideration arrive ultimately at the following result: Let AP be a tangent to the circle from A, and BR a tangent to the circle from B; then the swiftest course consists of the straight line AP, the arc PR, and the straight line RB.

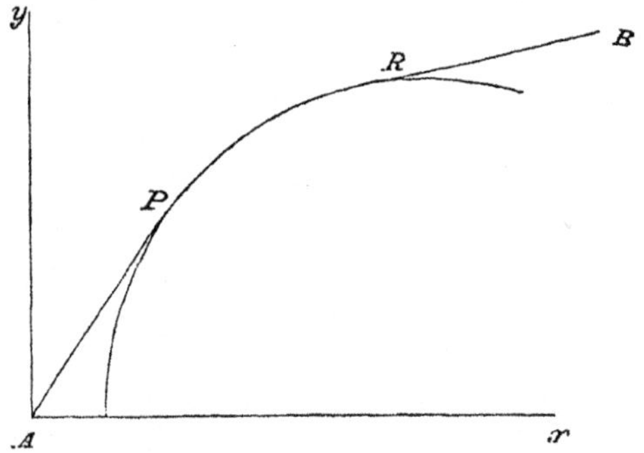

35. The solution here given furnishes a very good illustration of the important principle of Art. 18. Put

$$u = \int \phi(p)\,dx;$$

then to the first order

$$\delta u = \phi'(p)\,\delta y + \int \left\{ -\delta y \frac{d}{dx} \phi'(p) \right\} dx$$

$$= \phi'(p)\,\delta y - \int \delta y\, \phi''(p)\, y''\, dx.$$

Now along AP and RB we have $y'' = 0$, so that the part of δu under the integral sign reduces to only so much as arises from the elements between P and R. Now for such elements y'' is negative, and δy is essentially positive, so that this part of δu is positive. Since AP *touches* the circle the value of $\phi'(p)$ is the same at P whether we consider the straight line AP or the circle: thus the term $\phi'(p)\,\delta y$ will enter twice with equal value and opposite signs, and so vanishes. Similarly the term $\phi'(p)\,\delta y$ at R vanishes. Hence δu reduces to a small quantity of the first order which is essentially positive; and thus a minimum is secured.

I here suppose that in comparing the proposed path with an adjacent path we vary the *whole path:* if however we do not vary PR but only the pieces AP and RB the value of δu will still be positive, but it will be a small quantity of the *second* order instead of the *first*, namely, $\frac{1}{2}\int \phi''(p)\,(\delta p)^2\, dx$.

36. We may observe that with the law of velocity adopted in Art. 34 the swiftest course in Art. 32, if the direct course is forbidden, will consist of the straight lines AO and OB. In this case, proceeding as in Art. 35, we find that δu reduces to

$$\{3p_0\sqrt{(1+p_0^2)} - 3p_1\sqrt{(1+p_1^2)}\}\,\delta y,$$

where p_0 is the tangent of OAx, and p_1 the tangent of BOC; and δy is the variation of y at the point O. Now δy is essentially positive, and p_0 is by supposition greater than p_1: thus δu is a positive quantity of the first order.

37. It is obvious that the process of Arts. 34 and 35 does not require that the curve which bounds the forbidden area should necessarily be a circular arc; any curve which is concave to the

axis of x may be taken instead of the circular arc. Nor need the law of velocity be necessarily that expressed by $\phi(p) = \frac{1}{1+p^2}$: in general $\phi(p)$ may be any function of p such that $\phi'(p) = 0$ [or $\phi'(p) = C$] has no possible root or only one possible root.

38. As another example let the law of velocity be such that $\phi(p) = b^2 - \frac{a^2p^2}{2} + \frac{p^4}{4}$: see Art. 9. This in fact requires that the expression for the velocity should be $\dfrac{\sqrt{(1+p^2)}}{b^2 - \frac{a^2p^2}{2} + \frac{p^4}{4}}$.

Suppose A and B so situated that the swiftest course from A

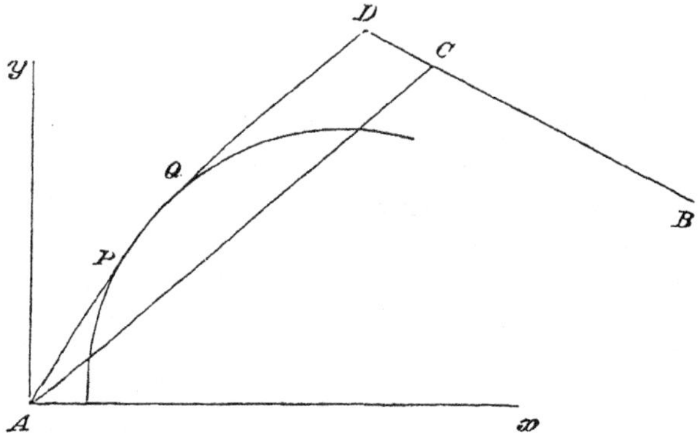

to B when there is no obstacle consists of the straight lines AC and CB which are determined by $p = \pm a$. And suppose that a certain circular area is not to be crossed so that this course cannot be adopted. Required the swiftest course.

It will be found that the swiftest course consists of the following parts: the straight line AP drawn from A to touch the circle; then the arc PQ where Q is such that the tangent at Q is parallel to AC; then the straight line QD which is part of this tangent; and finally the straight line DCB.

That we thus obtain a minimum can be shewn as in Art. 35. If we pass from the assigned path to an adjacent path the variation will be found to be a *positive* quantity: it will be of the *first* order if we vary the whole path, and of the *second* order if we vary all except the part PQ.

39. In the problems of Arts. 34 and 38 it may be granted that we have obtained *minima* results, but yet it may be asked, how do we know that we have obtained the *least* results? I answer that the Calculus of Variations is immediately concerned only with maxima and minima values; and it would have been sufficient to use the term minimum in these problems. Nevertheless we can see that there must be a least value in each problem, and that the path must consist of a straight line or lines with perhaps part of the boundary of the forbidden area. Then, on trying various combinations of these possible components, we may soon convince ourselves that there can be no least values except those which have been assigned.

The kind of discontinuity which these problems furnish is that of straight lines and a curve which *touch* where they meet.

40. Suppose we seek for a curve of maximum or minimum length between two given points.

Let $u = \int \sqrt{(1+p^2)}\, dx$; then as far as terms of the second order inclusive,

$$\delta u = \int \left\{ \frac{p\delta p}{\sqrt{(1+p^2)}} + \frac{(\delta p)^2}{2(1+p^2)^{\frac{3}{2}}} \right\} dx$$

$$= \frac{p\delta y}{\sqrt{(1+p^2)}} - \int \frac{d}{dx}\left(\frac{p}{\sqrt{1+p^2}}\right) \delta y\, dx + \frac{1}{2}\int \frac{(\delta p)^2\, dx}{(1+p^2)^{\frac{3}{2}}}.$$

Thus in the usual way we obtain

$$\frac{p}{\sqrt{(1+p^2)}} = \text{constant},$$

and this corresponds to a minimum.

As long as we impose no other condition there will be no *maximum*.

But now let us propose to find a line of maximum length between two fixed points with the condition that y is always positive and that $\frac{dy}{dx}$ *never changes sign.*

Let A and B be the two fixed points; take A for the origin,

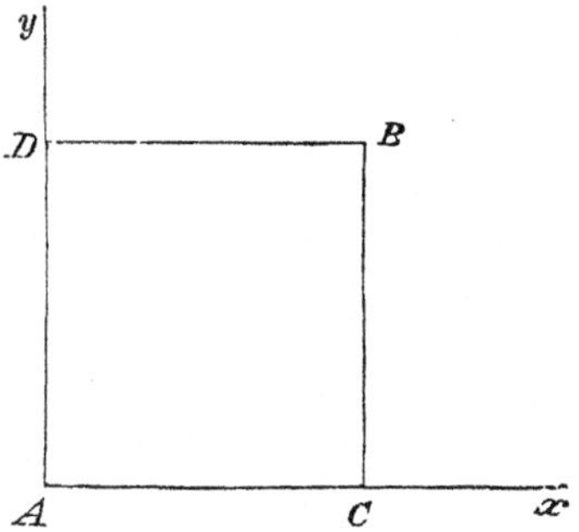

and draw BC and BD perpendiculars on the axes. Then the conditions imposed assign AD and DB as forming one boundary which the required line must not transgress, and AC and CB as another boundary.

The preceding investigation shews that there will be no maximum whatever so long as we take any line except one of these imposed boundaries.

41. It remains to investigate whether these boundaries themselves are the required lines of maximum length. In order to avoid infinite values of p we will transform to polar co-ordinates; the origin may be supposed to be at C.

Let $u = \int \sqrt{\left\{r^2 + \left(\frac{dr}{d\theta}\right)^2\right\}}\, d\theta$; then to the first order

$$\delta u = \frac{r'\delta r}{\sqrt{(r^2 + r'^2)}} + \int \delta r \left[\frac{r}{\sqrt{(r^2 + r'^2)}} - \frac{d}{d\theta}\frac{r'}{\sqrt{(r^2 + r'^2)}}\right] d\theta,$$

where r' is put for $\frac{dr}{d\theta}$.

Now the expression $\frac{r}{\sqrt{(r^2 + r'^2)}} - \frac{d}{d\theta}\frac{r'}{\sqrt{(r^2 + r'^2)}}$ is zero both along AD and along DB; thus the part of δu which is under the integral sign vanishes.

The other part of δu does not vanish. For the straight line AD we have $r = \frac{a}{\cos\theta}$, where $AC = a$: thus $\frac{r'}{\sqrt{(r^2 + r'^2)}} = \sin\theta$. Similarly for the straight line DB we have $r = \frac{b}{\sin\theta}$, where $AD = b$:

thus $\frac{r'}{\sqrt{(r^2 + r'^2)}} = -\cos\theta$.

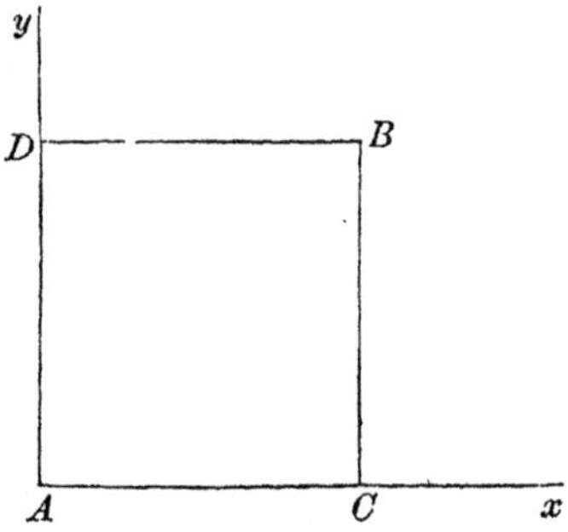

Hence finally

$$\delta u = (\sin DCA + \cos DCA)\,\delta r$$

where δr corresponds to the point D. And as when we pass by variation from the boundary ADB to an adjacent curve we must keep within the boundary, δr is essentially negative. Thus δu is necessarily negative and our result is a maximum.

[If however δr corresponding to the point D is zero, then δu is of the second order instead of the first order, and is positive; but then $\frac{dy}{dx}$ does not retain the same sign throughout.]

In like manner the boundary ACB constitutes a maximum.

42. It is required to draw a curve of given length between two fixed points so that the area bounded by the curve and the straight line joining the fixed points may be a maximum.

This is a well-known problem; the curve must be an arc of a circle: see Todhunter's *History of the Calculus of Variations*, page 69. But suppose we impose the condition that the curve is not to pass beyond a certain given straight line which contains one of the fixed points.

Let A and B be the fixed points, and OA a fixed straight line

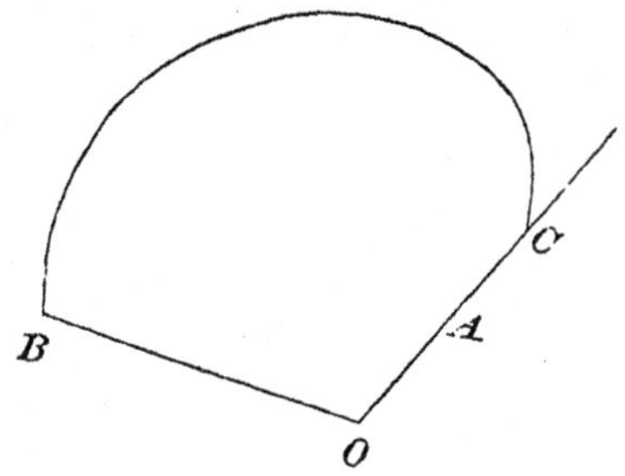

which the required curve is not to transgress. Take any point O in this fixed straight line as the origin of polar co-ordinates. As the curve is not to transgress the fixed straight line, the most general supposition we can make is that it must consist of some portion AC of this fixed straight line, of length at present unknown, and of some curve CB.

Let $AC = r_0$, and let the angle $BOA = \beta$. Then we have to find the maximum of $\frac{1}{2}\int_0^\beta r^2 d\theta$ while $r_0 + \int_0^\beta \sqrt{r^2 + \left(\frac{dr}{d\theta}\right)^2} d\theta$ has a given value. Hence a denoting a constant, we have by the usual theory to seek the maximum of

$$ar_0 + \int_0^\beta \left\{\frac{1}{2} r^2 + a\sqrt{r^2 + \left(\frac{dr}{d\theta}\right)^2}\right\} d\theta.$$

Denote this by u. Then in the usual way we make the part of δu which is under the integral sign vanish: and thus we find that the curve must be a circular arc. Then we have left

$$\delta u = a\delta r_0 \left\{1 - \frac{\frac{dr}{d\theta}}{\sqrt{r^2 + \left(\frac{dr}{d\theta}\right)^2}}\right\}_0.$$

To make this vanish we must have $\left(\frac{dr}{d\theta}\right)_0$ infinite, that is $\left(\frac{d\theta}{dr}\right)_0 = 0$; thus the circular arc must *touch* the fixed straight line. Then r_0 must be determined from this condition, and the circumstance that the whole perimeter is a given quantity.

43. For a more general problem we may take instead of a fixed *straight line* passing through A a fixed *curve;* and impose the condition that the required curve shall not pass beyond this fixed curve. It is natural to conjecture that if the given perimeter is so large that the required curve consists in part of an arc of the fixed curve, then the circular arc will *touch* the fixed curve at the point of junction. This we shall now verify.

Suppose the required curve to consist of AP a portion of the fixed curve APC, and the arc PB which we know will be an arc of a circle. Take any point O in AB as origin of polar co-ordinates. Let the unknown angle AOP be denoted by γ. Put l for AP and S for the area AOP. Then a denoting a constant, by the

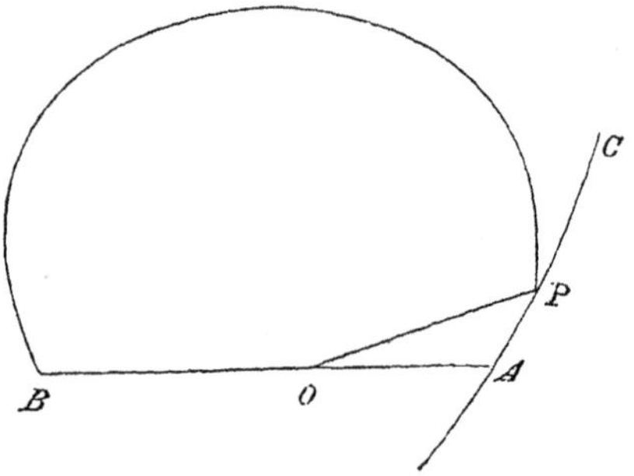

usual theory we must investigate the maximum of

$$al + S + \int_\gamma^\pi \left\{\frac{r^2}{2} + a\sqrt{(r^2 + r'^2)}\right\} d\theta.$$

Denote this by u. Then by considering the part of δu which is under the integral sign and making it vanish, we obtain a circular arc for the curve PB. Thus we have left

$$\delta u = a\delta l + \delta S - \left\{\frac{r^2}{2} + a\sqrt{(r^2 + r'^2)}\right\} d\theta - \frac{ar'\delta r}{\sqrt{(r^2+r'^2)}},$$

where for θ we are to put γ.

Now suppose that $r = \psi(\theta)$ is the equation to the fixed curve; then $\delta l = \sqrt{r^2 + \{\psi'(\theta)\}^2}\, d\theta$. Also $\delta S = \frac{r^2}{2} d\theta$. Thus

$$\delta u = a\sqrt{r^2 + \{\psi'(\theta)\}^2}\, d\theta - a\sqrt{(r^2 + r'^2)}\, d\theta - \frac{ar'\delta r}{\sqrt{(r^2+r'^2)}},$$

where for θ we are to put γ.

But since the point P must be on the fixed curve we obtain [by a process like that in Todhunter's *Integral Calculus*, third edition, Art. 359,] the condition

$$\delta r = \{\psi'(\theta) - r'\}\, d\theta;$$

and thus

$$\delta u = a\left\{\sqrt{r^2 + \{\psi'(\theta)\}^2} - \frac{r^2 + r'\psi'(\theta)}{\sqrt{(r^2+r'^2)}}\right\} d\theta.$$

Hence that this may vanish we must have when $\theta = \gamma$

$$\sqrt{r^2 + \{\psi'(\theta)\}^2} \times \sqrt{r^2 + r'^2} = r^2 + r'\psi'(\theta);$$

this leads to $\{r' - \psi'(\theta)\}^2 = 0$, so that $r' = \psi'(\theta)$.

Thus the statement is established.

44. In like manner we may treat the problem in which there is also a fixed curve passing through B which the required curve must not transgress. As an example we may refer to the special case originally discussed by Legendre where the boundaries which the required curve must not transgress consist of two parallel straight lines, one passing through one of the fixed points, and the other through the other fixed point.

45. An area is to be bounded by a perimeter of given length; the perimeter is to be constrained to pass through a certain number of fixed points: determine the form of the figure so that the area may be a maximum.

[This problem was proposed by the present writer in the Mathematical Tripos Examination of 1865.]

Suppose for simplicity of conception that there are *three* fixed points. Take an origin of polar co-ordinates within the triangle formed by joining the points. Then a denoting a constant we have by the usual theory to find the maximum of

$$\int_0^{2\pi} \left\{ \frac{r^2}{2} + a\sqrt{r^2 + \left(\frac{dr}{d\theta}\right)^2} \right\} d\theta,$$

with the condition that for three assigned values of θ the values of r must be equal to certain known quantities.

Hence we shall obtain for the required curve three arcs of circles all having the *same radius*, namely a.

Thus we have the discontinuity of a curve composed of arcs which meet at a finite angle.

46. But now let us advert to the case in which the given perimeter is so long that the arcs of circles would intersect outside the triangle formed by joining the three fixed points. In this case some portions of the area would in fact be counted twice over. If however we reject this as inconsistent with the nature of the problem, we must consider what modifications we have to make in our solution.

From the discussion in Art. 42 we are now led to the following conclusion:

Let A, B, C be the given points; then the required figure will be composed of certain straight lines AA', BB', CC', and of certain circular arcs $A'B'$, $B'C'$, $C'A'$, all having the same radius. Moreover the circular arcs will *touch* the straight lines which they respectively meet.

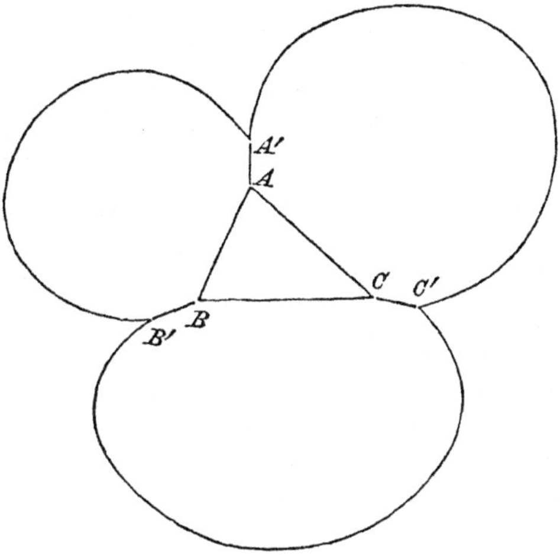

47. We may give a mechanical aspect to this problem.

Suppose a cylindrical vessel to be formed of flexible material, and placed on a horizontal plane; and suppose that the material is constrained to pass round fixed vertical rods. Let fluid be poured in; then we know that in the state of stable equilibrium the area of the base will have a maximum value, so that the centre of gravity may be as low as possible.

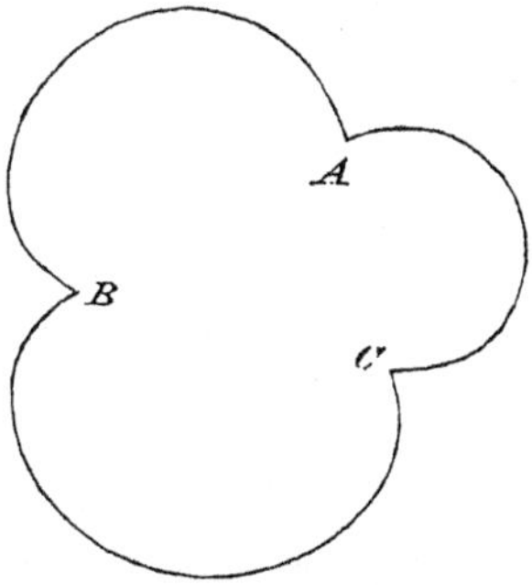

We know from almost elementary considerations that each portion of the boundary of the base will be an arc of a circle. The fact that the radii are all equal may be deduced from the relation which connects the pressure and tension with the radius: the relation is given in books on Hydrostatics.

Then, if the perimeter of the flexible material is large enough, we obtain the result explained in Art. 46. The lengths of the straight portions AA', BB', CC' will of course be counted twice.

48. The discussion of the lengths and the positions of the rectilinear parts of the figure is not a problem of the Calculus of Variations, but of ordinary Geometry and Differential Calculus: we will therefore offer only a few remarks on it.

49. We shall shew that as the perimeter is gradually increased a rectilineal portion occurs first at the *largest* angle of the triangle ABC.

Suppose the arcs which have AB and AC as chords to touch

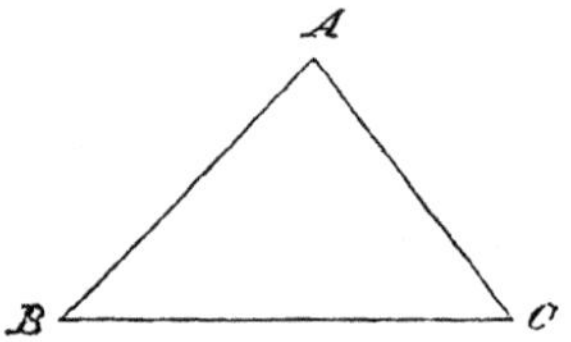

at A. Let r denote the radius of each arc; and let a, b, c denote the sides of the triangle.

Then $$A + \cos^{-1}\frac{c}{2r} + \cos^{-1}\frac{b}{2r} = \pi \text{ .... (1).}$$

If instead of touching at A the arcs which have B as a common point touched we should have

$$B + \cos^{-1}\frac{c}{2r} + \cos^{-1}\frac{a}{2r} = \pi \text{ .... (2).}$$

Now if A is greater than B, then $\cos^{-1}\frac{b}{2r}$ is greater than $\cos^{-1}\frac{a}{2r}$. This shews that as the perimeter is gradually increased the arcs touch first at the largest angle.

[Because as r gradually increases we arrive at the value which corresponds to (1) *before* we arrive at the value which corresponds to (2).]

50. Now suppose that there is only one rectilinear part of the figure, namely AA'.

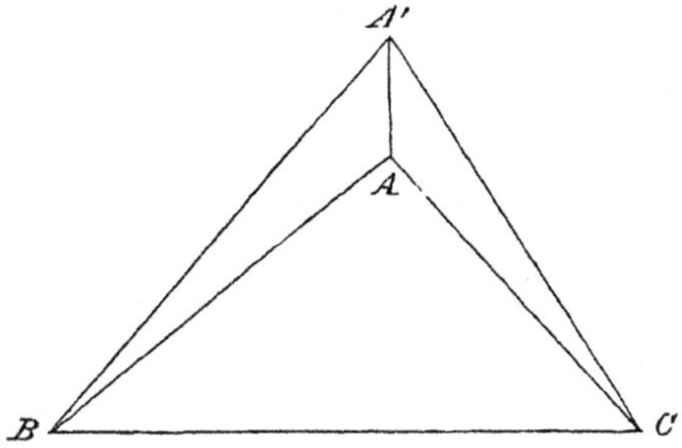

Let $$AA' = \rho, \qquad BAA' = \theta;$$

$$\frac{\sin BA'A}{\sin \theta} = \frac{c}{BA'}, \quad \frac{\sin CA'A}{\sin (2\pi - A - \theta)} = \frac{b}{CA'};$$

therefore $$\frac{\sin BA'A}{\sin CA'A} = \frac{CA'}{BA'} \cdot \frac{c \sin \theta}{b \sin (2\pi - A - \theta)}.$$

And $$2r \sin BA'A = BA', \quad 2r \sin CA'A = CA';$$

therefore $$\frac{c \sin \theta}{b \sin (2\pi - A - \theta)} = \left(\frac{BA'}{CA'}\right)^2 = \frac{c^2 + \rho^2 - 2c\rho \cos \theta}{b^2 + \rho^2 - 2b\rho \cos (2\pi - A - \theta)}.$$

This gives a curve of the third degree as the locus of A'.

If we put $\rho = 0$ we get $\dfrac{\sin \theta}{\sin (2\pi - A - \theta)} = \dfrac{c}{b}$,

this determines the direction which AA' initially takes.

If we denote by D the middle point of BC we shall find that initially $\theta = \pi - CAD$. For denoting CAA' by ϕ we have

$$\frac{\sin (\pi - \theta)}{\sin (\pi - \phi)} = \frac{\sin C}{\sin B}, \text{ and } \pi - \theta + \pi - \phi = A;$$

also $$\frac{\sin CAD}{\sin BAD} = \frac{\sin C}{\sin B}, \text{ and } CAD + BAD = A.$$

51. Next consider the case in which there are *two* rectilinear parts of the figure, namely AA' and BB'.

Thus two circular arcs described on CA' and $B'A'$ as chords with the same radius touch at A'; and similarly those described on CB' and $B'A'$ touch at B'.

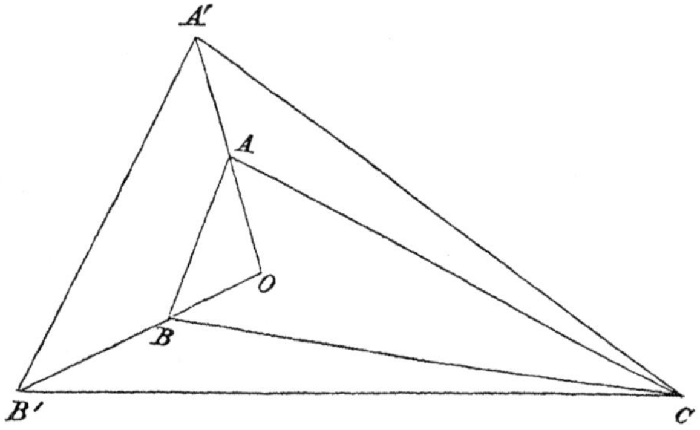

It follows by applying such equations as (1) and (2) of Art. 49 to the triangle $CA'B'$ that $CB' = CA'$. Let $A'A$ and $B'B$ meet at O.

Then $A'O = B'O$; for they are tangents to the same circle. Hence the angle $CB'B =$ the angle $CA'A$.

Put $AA' = \rho$, $BAA' = \theta$, $BB' = \rho'$, $B'BA = \theta'$.

Now $$\frac{\sin CA'A}{\sin CAA'} = \frac{CA}{CA'}; \text{ therefore}$$

$$\sin CA'A = \frac{CA}{CA'} \sin (2\pi - A - \theta).$$

Similarly

$$\sin CB'B = \frac{CB}{CB'} \sin (2\pi - B - \theta');$$

therefore $$\frac{\sin (A + \theta)}{CB} = \frac{\sin (B + \theta')}{CA},$$

that is $$\frac{\sin (A + \theta)}{\sin A} = \frac{\sin (B + \theta')}{\sin B} \quad \text{.........................(1).}$$

Also $OA' = OB'$; that is

$$-\frac{c \sin \theta}{\sin (\theta + \theta')} + \rho = -\frac{c \sin \theta'}{\sin (\theta + \theta')} + \rho' \ldots\ldots\ldots\ldots\ldots\ldots (2).$$

And as $CA' = CB'$ we have

$$b^2 + \rho^2 - 2b\rho \cos (A + \theta) = c^2 + \rho'^2 - 2c\rho' \cos (B + \theta') \ldots\ldots (3).$$

The equations (1), (2), and (3) will theoretically furnish by elimination one relation between ρ and θ, and one relation between ρ' and θ'.

52. In the case in which the perimeter is so large that there are three rectilinear parts of the figure, the result is very simple.

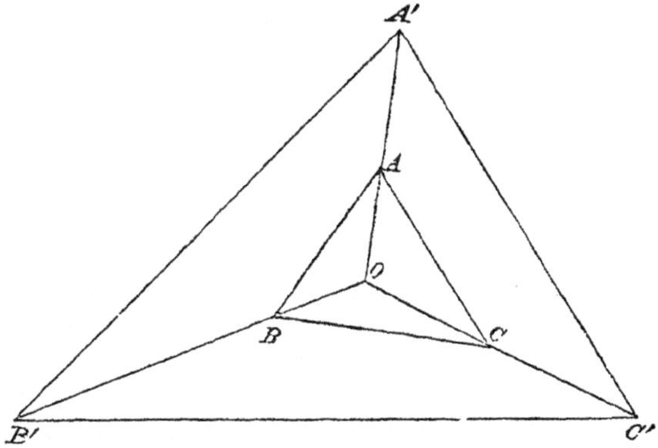

Let AA', BB', CC' denote these rectilinear parts. The triangle $A'B'C'$ will be equilateral. The three straight lines $A'A$, $B'B$, $C'C$ being produced will meet at a point O, so that the angles AOB, BOC, COA will all be angles of 120°. Thus O will be a fixed point in the triangle ABC.

If the triangle ABC has an angle greater than 120° suppose it to be the angle A; then the fixed point O will be outside the triangle ABC, and BOC will be an angle of 120° while BOA and COA will each be an angle of 60°.

53. Suppose it required to find the greatest area included within a given figure, and having a perimeter of given length.

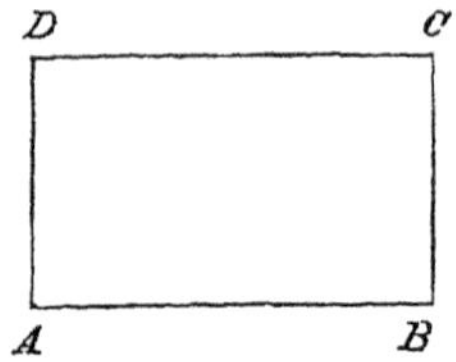

For simplicity let us take the given figure to be a rectangle $ABCD$; let AD be the shorter side.

I. If the length of the given perimeter does not exceed πAD the required solution is of course a circle.

II. If the length of the given perimeter exceeds πAD the required solution cannot be a circle; for a circle with a perimeter of the given length would not fall entirely within the rectangle. But we are sure that the perimeter cannot consist of anything but some combination of circular arcs with parts of the boundary of the rectangle. For as in Art. 17 we know that in general it is necessary that the part of the variation under the integral sign must vanish, the only exception being when, as at the boundary of the rectangle, the quantity denoted by δy cannot take either sign.

Hence in the present problem the required solution must consist of a combination of straight lines and arcs of circles. By

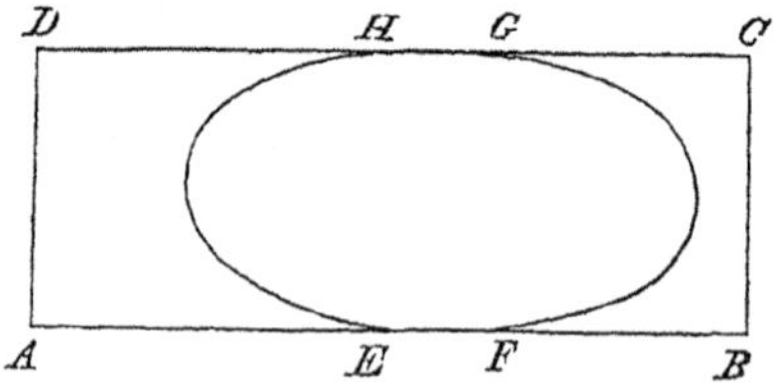

considerations similar to those in Art. 42 we infer that the straight lines will *touch* the arcs. Thus we obtain two straight lines EF

and GH, and two semicircular arcs FG and HE; the lengths of EF and GH being of course determined by the condition that the whole perimeter shall have a given length.

This solution holds provided the given length does not exceed $\pi AD + 2AB - 2AD$.

III. If the given length exceeds $\pi AD + 2AB - 2AD$ the required solution consists of four straight lines, EF, GH, KL, MN

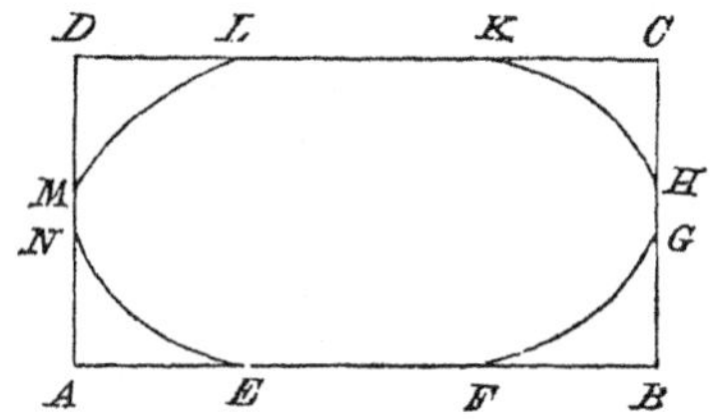

connected by four quadrants of a circle, FG, HK, LM, NE, [of the same radius].

54. Required to find the shortest path from a fixed point A to a fixed point B, supposing a circular obstacle having its centre in the straight line AB; and supposing that the path is never to be convex towards AB and to have its radius of curvature never less than a given quantity r, and to have no abrupt change of direction.

I. If the radius of the obstacle is not less than r the path

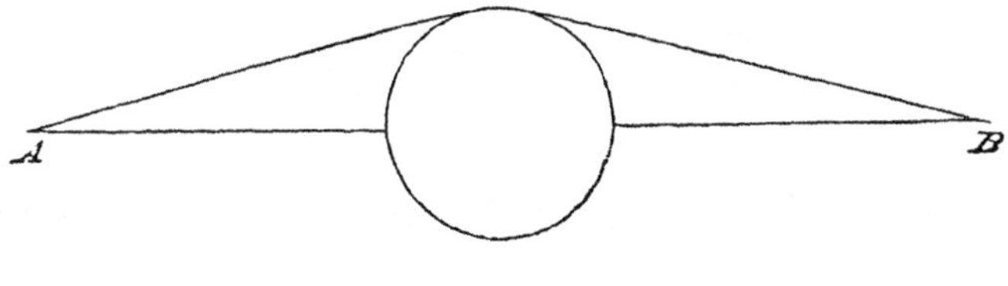

consists of two tangents to the obstacle with an arc of the obstacle.

II. If the radius of the obstacle is less than r the path consists of an arc of radius r which touches the obstacle, and straight

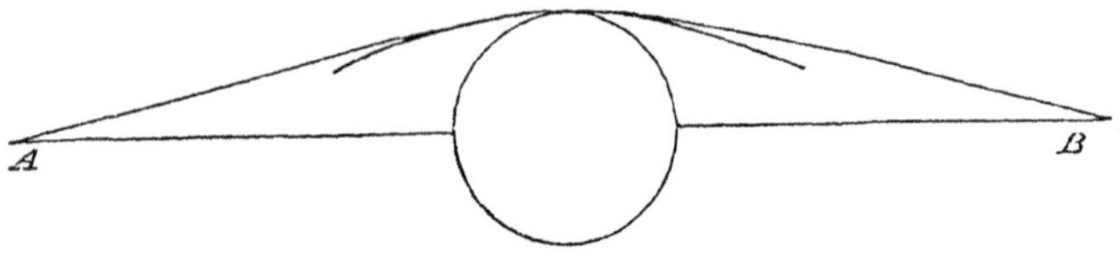

lines from A and B which are tangents to this arc. If the obstacle is midway between A and B the two tangents will be of equal length, the arc touching the obstacle at the point most distant from AB; but if the obstacle is not midway between A and B the point of contact of the arc and the obstacle must be found by the Differential Calculus.

III. The solution holds so long as the arc which touches the obstacle cuts AB at two points between A and B. If this cannot be secured we must describe a circle on AB as chord with radius r: then if the shorter arc will not clear the obstacle the longer arc must be taken.

If however we are not to transgress the limits obtained by drawing straight lines through A and B at right angles to AB, it will in this case be impossible to find *any* path that satisfies the conditions, and so of course there can be no *shortest* path.

55. It is easy to justify the statements in the preceding Article by the Calculus of Variations.

Let $u = \int \sqrt{(1+p^2)}\, dx$; so that when the integral is taken between proper limits u denotes the length of the path. Then to the second order inclusive

$$\delta u = \frac{p\delta y}{\sqrt{(1+p^2)}} - \int \frac{d}{dx}\left(\frac{p}{\sqrt{(1+p^2)}}\right) \delta y\, dx + \frac{1}{2}\int \frac{(\delta p)^2\, dx}{(1+p^2)^{\frac{3}{2}}}.$$

The term outside the integral sign vanishes since the extreme points A and B are fixed. Thus we have

$$\delta u = -\int \frac{y''\delta y\, dx}{(1+p^2)^{\frac{3}{2}}} + \frac{1}{2}\int \frac{(\delta p)^2\, dx}{(1+p^2)^{\frac{3}{2}}}.$$

Now δu is positive. For along the straight lines we have $y''=0$, so that δu reduces to the second part of the above expression which is a positive term of the second order. And along the circular arc we have y'' negative, and it will [probably] be found that we cannot suppose δy negative without breaking the conditions which have been imposed, that the curve is not to be convex to the axis, and that the radius of curvature is not to be less than r, and that there is to be no abrupt change of direction. Thus we see that we have a minimum, and we infer that it is the *least* value because no other presents itself.

56. We have thus briefly considered various simple examples of discontinuous solutions; we shall now proceed to a full discussion of some problems of historical interest in the Calculus of Variations, which involve discontinuity.

CHAPTER IV.

MINIMUM SURFACE OF REVOLUTION.

57. REQUIRED the plane curve joining two given points which by revolving round a given axis in its plane will generate a surface of minimum area.

This problem has been much discussed: see the prize essay by Goldschmidt noticed in Todhunter's *History of the Calculus of Variations*, page 340; see also Professor Jellett's *Calculus of Variations*, 1850, page 145, the *Calcul des Variations*, published in 1861 by Moigno and Lindelöf, page 204, and Dienger's *Grundriss der Variationsrechnung*, 1867, page 15. I shall add something to the researches of previous writers.

58. Take the axis of x as that of revolution. Then we require the minimum of $\int y \sqrt{(1+p^2)}\, dx$, the limiting values of x and y being fixed.

We obtain in the usual way

$$\frac{y}{\sqrt{(1+p^2)}} = C_1 \text{ a constant};$$

thus

$$p^2 = \frac{y^2 - C_1^2}{C_1^2};$$

therefore

$$\frac{dx}{dy} = \pm \frac{C_1}{\sqrt{(y^2 - C_1^2)}};$$

therefore

$$x + C_2 = \pm C_1 \log \{y + \sqrt{(y^2 - C_1^2)}\}.$$

Taking either sign we have the equation to a catenary of which the axis of x is the directrix. We have then to examine this solution.

59. First we ask if it is possible to draw a catenary having a given directrix, and passing through two given points. This question has been considered by previous writers; the conclusion is that sometimes two catenaries can be drawn, sometimes only one, and sometimes none. I shall however presently give a new investigation of the question, simpler I believe than those hitherto published.

The next question is to determine whether we really obtain a minimum. I shall shew that when two catenaries can be drawn the upper corresponds to a minimum, and the lower does not; and that when only one catenary can be drawn it does not correspond to a minimum. These statements are new. Goldschmidt erroneously thought that both catenaries corresponded to a minimum: see his page 27. The other writers do not discriminate between the two catenaries, except Moigno and Lindelöf in a particular case. Strauch is not very full on this problem; he considers that there is always a minimum corresponding to the catenary: see his Vol. II page 276. Stegmann does not discuss the problem: he barely alludes to it on his page 187.

60. It is convenient to begin with the particular case in which the given points are equally distant from the axis; to this particular case some previous writers have practically restricted themselves.

Let the distance of each given point from the axis be b; and let $2a$ be the distance between the points. Then we assume for the equation to the catenary

$$y = \frac{c}{2}\left(e^{\frac{x}{c}} + e^{-\frac{x}{c}}\right);$$

so that c has to be found from the equation

$$b = \frac{c}{2}\left(e^{\frac{a}{c}} + e^{-\frac{a}{c}}\right) \qquad (1).$$

We have then in fact to determine if the last equation gives a real value or real values for c.

Denote $\frac{c}{2}(e^{\frac{a}{c}}+e^{-\frac{a}{c}})$ by $\phi(c)$; regard a as a constant and c as a variable. Then we see that $\phi(c)$ is infinite both when $c=0$, and when $c=\infty$. And

$$2\phi'(c)=e^{\frac{a}{c}}+e^{-\frac{a}{c}}-\frac{a}{c}(e^{\frac{a}{c}}-e^{-\frac{a}{c}}) \text{ ....(2).}$$

From (2) we can shew that $\phi'(c)$ vanishes once, and only once, as c ranges from zero to infinity. For by expanding we get

$$\phi'(c)=1-\frac{1}{2}\frac{a^2}{c^2}-\frac{3}{\underline{|4}}\frac{a^4}{c^4}-\ldots-\frac{2n-1}{\underline{|2n}}\frac{a^{2n}}{c^{2n}}-\ldots;$$

so that $\phi'(c)$ is negative infinity when c is zero, and is unity when c is infinite, and changes sign once, and only once, as c passes from zero to infinity. And $\phi(c)$ has its least value when $\phi'(c)=0$.

If then the given value of b be greater than the least value of $\phi(c)$ there are two values of c which satisfy (1); if the given value of b be equal to the least value of $\phi(c)$ there is only one value of c; if the given value of b be less than the least value of $\phi(c)$ there is no possible value of c.

It has been found that the value of $\frac{a}{c}$ which makes

$$e^{\frac{a}{c}}+e^{-\frac{a}{c}}-\frac{a}{c}(e^{\frac{a}{c}}-e^{-\frac{a}{c}})=0,$$

is approximately $\frac{a}{c}=1\cdot19968\ldots$; and then it follows from (1) that $\frac{b}{c}=1\cdot81017\ldots$; and therefore $\frac{b}{a}=1\cdot5088\ldots$: see Dienger, and Moigno and Lindelöf. Thus there are two catenaries satisfying the prescribed conditions, or one, or none according as $\frac{b}{a}$ is greater than, equal to, or less than $1\cdot5088$.

61. Now we pass to the question whether corresponding to a catenary the surface generated is a minimum. We know that

to ensure a minimum the tangents to the catenary at the fixed points must intersect above the axis of x; see Art. 29.

From symmetry the two tangents in the present case will intersect on the axis of y.

The equation to the tangent to the catenary at the point (a, b) is

$$y - b = p\,(x - a),$$

where $p = \frac{1}{2}(e^{\frac{a}{c}} - e^{-\frac{a}{c}})$; therefore the ordinate of the point where this crosses the axis of y is $b - pa$, that is

$$\frac{c}{2}(e^{\frac{a}{c}} + e^{-\frac{a}{c}}) - \frac{a}{2}(e^{\frac{a}{c}} - e^{-\frac{a}{c}}).$$

And it is obvious from our discussion of the value of $\phi'(c)$ that the above expression is positive for the larger value of c obtained from (1) and negative for the smaller value of c. Hence when two catenaries can be drawn the upper catenary corresponds to a minimum and the lower does not. When only one catenary can be drawn it does not correspond to a minimum.

The result for this particular case had been obtained by Moigno and Lindelöf: see their page 210.

62. We shall now discuss the general problem.

Let b be the distance of one given point from the axis of x, and k the distance of the other.

Let the axis of y be placed midway between the given points; take a for the abscissa of the former given point, and $-a$ for the abscissa of the latter. Then we take for the equation to the catenary

$$y = \frac{c}{2}(e^{\frac{x+n}{c}} + e^{-\frac{x+n}{c}}),$$

where n and c have to be found from the equations

$$\left.\begin{aligned} b &= \frac{c}{2}(e^{\frac{a+n}{c}} + e^{-\frac{a+n}{c}}) \\ k &= \frac{c}{2}(e^{\frac{n-a}{c}} + e^{-\frac{n-a}{c}}) \end{aligned}\right\} \quad\ldots\ldots\ldots\ldots\ldots\ldots\ldots\ldots(3).$$

From (3) we obtain

$$\left.\begin{aligned}\frac{c}{2}e^{\frac{n}{c}}\left(e^{\frac{2a}{c}}-e^{-\frac{2a}{c}}\right)&=be^{\frac{a}{c}}-ke^{-\frac{a}{c}}\\ \frac{c}{2}e^{-\frac{n}{c}}\left(e^{-\frac{2a}{c}}-e^{\frac{2a}{c}}\right)&=be^{-\frac{a}{c}}-ke^{\frac{a}{c}}\end{aligned}\right\}\ \ldots\ldots\ldots\ldots\ldots\ldots(4).$$

And from (4) by multiplication

$$\frac{c^2}{4}\left(e^{\frac{2a}{c}}-e^{-\frac{2a}{c}}\right)^2=\left(be^{\frac{a}{c}}-ke^{-\frac{a}{c}}\right)\left(ke^{\frac{a}{c}}-be^{-\frac{a}{c}}\right)\ \ldots\ldots\ldots\ (5).$$

Thus we have eliminated n and obtained the equation (5) for determining c: we have now to examine if this equation gives a real value or real values for c.

Let $\phi(c)$ stand for

$$\frac{c^2}{4}\left(e^{\frac{2a}{c}}-e^{-\frac{2a}{c}}\right)^2-\left(be^{\frac{a}{c}}-ke^{-\frac{a}{c}}\right)\left(ke^{\frac{a}{c}}-be^{-\frac{a}{c}}\right).$$

Then $\phi(c)$ is infinite when c is zero, and is $4a^2+(b-k)^2$ when c is infinite.

We shall find that $\phi'(c)$

$$=\left(e^{\frac{2a}{c}}-e^{-\frac{2a}{c}}\right)\left\{\frac{c}{2}\left(e^{\frac{2a}{c}}-e^{-\frac{2a}{c}}\right)-a\left(e^{\frac{2a}{c}}+e^{-\frac{2a}{c}}\right)+\frac{2abk}{c^2}\right\}.$$

The factor $e^{\frac{2a}{c}}-e^{-\frac{2a}{c}}$ cannot change sign. The other factor of $\phi'(c)$ becomes by expansion

$$2a\left\{\frac{bk-\frac{4a^2}{3}}{c^2}-\frac{2^4a^4}{c^4}\left(\frac{1}{\lfloor 4}-\frac{1}{\lfloor 5}\right)-\ldots-\frac{2^{2n}a^{2n}}{c^{2n}}\left(\frac{1}{\lfloor 2n}-\frac{1}{\lfloor 2n+1}\right)-\ldots\right\}.$$

If bk is not greater than $\frac{4a^2}{3}$ then $\phi'(c)$ never changes its sign; and the least value of $\phi(c)$ is when c is infinite: as this value is positive it follows that $\phi(c)$ cannot vanish. If bk is greater than $\frac{4a^2}{3}$ then $\phi'(c)$ changes its sign once, and only once; so that $\phi(c)$ has a corresponding minimum value, and according as this value is negative, zero, or positive we have from (5) two values of c, or one, or none.

63. We now pass to the question whether corresponding to a catenary the surface generated is a minimum. As before we must determine whether the tangents to the catenary at the fixed points do or do not intersect above the axis of x.

Let p_1 stand for the value of $\frac{dy}{dx}$ at the point (a, b); and p_2 for the value of $\frac{dy}{dx}$ at the point $(-a, k)$: then the equations to the tangents are respectively

$$y - b = p_1(x - a), \quad y - k = p_2(x + a);$$

at the point of intersection

$$\frac{y - b + p_1 a}{y - k - p_2 a} = \frac{p_1}{p_2};$$

therefore
$$y = \frac{2ap_1p_2 + kp_1 - bp_2}{p_1 - p_2}.$$

By using (4) we obtain for the equation to the catenary

$$y = \frac{1}{\lambda}\left\{e^{\frac{x}{c}}\left(be^{\frac{a}{c}} - ke^{-\frac{a}{c}}\right) + e^{-\frac{x}{c}}\left(ke^{\frac{a}{c}} - be^{-\frac{a}{c}}\right)\right\},$$

where λ stands for $e^{\frac{2a}{c}} - e^{-\frac{2a}{c}}$.

Hence we shall find that

$$p_1 = \frac{\mu b - 2k}{\lambda c}, \quad p_2 = \frac{2b - \mu k}{\lambda c},$$

where μ stands for $e^{\frac{2a}{c}} + e^{-\frac{2a}{c}}$.

Thus $p_1 - p_2 = \frac{(\mu - 2)(b + k)}{\lambda c}$ which is positive.

We have now to examine the sign of $2ap_1p_2 + kp_1 - bp_2$.

From the values of p_1 and p_2 we obtain

$$kp_1 - bp_2 = \frac{2\mu bk - 2(b^2 + k^2)}{\lambda c},$$

$$p_1p_2 = \frac{2\mu(b^2 + k^2) - (4 + \mu^2)bk}{\lambda^2c^2}.$$

Thus

$$2ap_1p_2 + kp_1 - bp_2 = \frac{2}{\lambda^2 c^2}\{(b^2 + k^2 - \mu bk)(2\mu a - \lambda c) + (\mu^2 - 4)\, abk\}.$$

But equation (5) may be written

$$\frac{c^2\lambda^2}{4} = \mu bk - (b^2 + k^2),$$

and

$$\mu^2 - 4 = \lambda^2,$$

so that

$$2ap_1p_2 + kp_1 - bp_2 = \frac{\lambda c}{2} - \mu a + \frac{2abk}{c^2}.$$

Now from our investigation of the value of $\phi'(c)$ in Art. 62 it follows that when there are two admissible values of c the expression $\frac{\lambda c}{2} - \mu a + \frac{2abk}{c^2}$ is positive for the greater value of c and negative for the less; and when there is only one admissible value this expression is zero. Hence when two catenaries can be drawn the upper corresponds to a minimum, and the lower does not. When only one catenary can be drawn it does not correspond to a minimum.

We may remark that the two catenaries which present themselves in this problem correspond to the figure which a uniform endless string will assume when hung over two pegs.

64. We shall now consider a discontinuous minimum which always exists. This has been noticed by no writer, I think, except

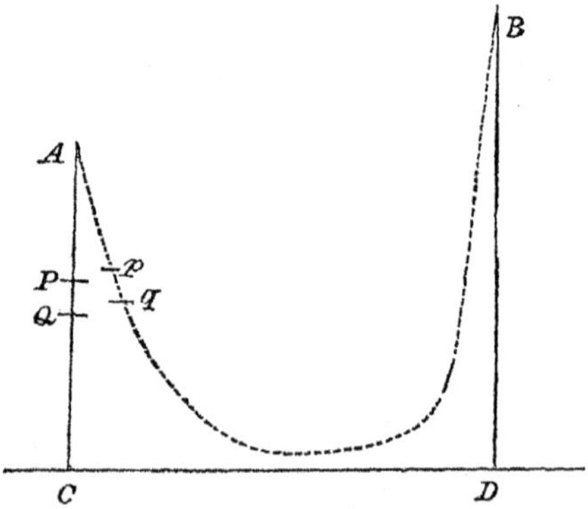

Goldschmidt; he briefly adverts to it, but does not shew that it is a minimum.

Let A and B be the given points; AC and BD the perpendiculars from them on the axis. Then the discontinuous solution is furnished by taking the generating curve to consist of AC, CD, and DB; so that the surface consists of the two circles having CA and DB respectively for their radii, connected by the straight line CD: this connecting part may be conceived to be an infinitesimally slender cylinder.

We shall hereafter consider the origin of this discontinuous solution: see Art. 68.

65. It is very easy to shew that the proposed solution really gives a surface of minimum area. Let the dotted line in the diagram represent a *closely adjacent* curve. Set off from A along AC and along the dotted line equal infinitesimal lengths. Let PQ and pq be a corresponding pair. Then it is plain that PQ will be rather nearer to the axis of revolution than pq is; and so pq will generate a somewhat larger element of area than PQ will. In like manner if we set off from B along BD and along the dotted line equal infinitesimal lengths we find that the element of the dotted line generates a somewhat larger element of area than the element of BD does. And CD generates no area, while any element of the dotted line in the neighbourhood of CD does generate an area since its distance from CD is not absolutely zero. Hence we see that the dotted line generates an area which is *certainly greater* than that of the proposed discontinuous solution; in other words the discontinuous solution really is a minimum.

66. The same result may be obtained by the ordinary methods of the Calculus of Variations.

In order to avoid infinite quantities we will use polar co-ordinates. Suppose the initial line parallel to DC; let k be the distance of the origin from DC; and let the vectorial angle increase in the direction from A towards B. Then the integral we wish to make a minimum is

$$\int (k - r\sin\theta) \sqrt{r^2 + \left(\frac{dr}{d\theta}\right)^2}\, d\theta,$$

the limiting values of the variables being fixed. Denote this integral by u; and put v for $\sqrt{r^2+\left(\frac{dr}{d\theta}\right)^2}$; also put p for $\frac{dr}{d\theta}$.

Then to the first order

$$\delta u = \int\left\{-v\sin\theta\,\delta r + \frac{k-r\sin\theta}{v}\,r\,\delta r + \frac{(k-r\sin\theta)\,p\,\delta p}{v}\right\}d\theta.$$

This transforms into

$$\frac{k-r\sin\theta}{v}p\,\delta r + \int\left\{-v\sin\theta + \frac{k-r\sin\theta}{v}r - \frac{d}{d\theta}\frac{(k-r\sin\theta)\,p}{v}\right\}\delta r\,d\theta.$$

The term outside the integral sign vanishes, for $\delta r = 0$ at the points A and B; and although δr is not zero at the points C and D where we have discontinuity, yet at these points $k - r\sin\theta = 0$.

We have now to determine whether the term in δu which is under the integral sign vanishes for every point of the discontinuous line which is under examination.

Consider the part CD; here $k - r\sin\theta = 0$, so that the coefficient of δr reduces to

$$-v\sin\theta + \frac{p\,(p\sin\theta + r\cos\theta)}{v},$$

that is to

$$\frac{pr\cos\theta - r^2\sin\theta}{v}.$$

Now when $k = r\sin\theta$ we get $p = -r\cot\theta$, so that the coefficient of δr is $-\frac{k^2}{v\sin\theta}$, which is a negative quantity and not zero; but along CD we have δr necessarily negative, since there is obviously the implied restriction that the generating curve shall be *above* the axis of x, or, which is the same thing, that the surface generated shall be taken positively. Hence as δr is negative δu is positive so far as the elements arising in connexion with CD are concerned.

Next consider the part AC; here we have $r\cos\theta =$ a constant $= l$ say. This gives $p = \frac{l\sin\theta}{\cos^2\theta}$, $v = \frac{l}{\cos^2\theta}$. Thus the coefficient of δr becomes

$$-\frac{l\sin\theta}{\cos^2\theta}+k\cos\theta-l\sin\theta-\frac{d}{d\theta}\frac{(k\cos\theta-l\sin\theta)\sin\theta}{\cos\theta}$$

$$=\frac{-l\sin\theta}{\cos^2\theta}-l\sin\theta+\frac{d}{d\theta}\frac{l(1-\cos^2\theta)}{\cos\theta}=0.$$

Similarly δu is zero so far as the elements arising in connexion with BD are concerned. Thus on the whole we have δu a positive quantity of the first order, and so the proposed solution is really a minimum.

67. Thus so long as we consider a curve which is adjacent to the discontinuous line but which differs from it through the whole extent, we are sure of a minimum without examining the terms of the *second* order in the variation. But if we do *not* vary the part CD our conclusion that δu is a positive quantity of the *first* order does not hold: so that we are interested in examining the terms of the second order in δu.

These terms consist of

$$\frac{1}{2}\int\left\{\left[\frac{-2r\sin\theta}{v}+\frac{(k-r\sin\theta)p^2}{v^3}\right](\delta r)^2\right.$$

$$+\left[\frac{-2p\sin\theta}{v}-\frac{2(k-r\sin\theta)rp}{v^3}\right]\delta r\,\delta p$$

$$\left.+\frac{(k-r\sin\theta)r^2}{v^3}(\delta p)^2\right\}d\theta$$

$$=-\int\left\{\frac{r\sin\theta}{v}(\delta r)^2+\frac{p\sin\theta}{v}\delta r\,\delta p\right\}d\theta$$

$$+\frac{1}{2}\int\frac{(k-r\sin\theta)(p\,\delta r-r\,\delta p)^2}{v^3}\,d\theta.$$

The second of these two terms is never negative. We will transform the first term. We have

$$\int\frac{p\sin\theta}{v}\delta r\,\delta p\,d\theta=(\delta r)^2\frac{p\sin\theta}{v}-\int\delta r\frac{d}{d\theta}\left(\frac{p\sin\theta\,\delta r}{v}\right)d\theta,$$

therefore

$$2\int \frac{p \sin\theta}{v}\,\delta r\,\delta p\,d\theta = (\delta r)^2 \frac{p\sin\theta}{v} - \int (\delta r)^2 \frac{d}{d\theta}\left(\frac{p\sin\theta}{v}\right) d\theta.$$

Thus the first term becomes

$$-(\delta r)^2 \frac{p\sin\theta}{2v} - \int \left\{ \frac{r\sin\theta}{v} - \frac{1}{2}\frac{d}{d\theta}\left(\frac{p\sin\theta}{v}\right)\right\} (\delta r)^2\,d\theta.$$

Now for AC the expression $\frac{r\sin\theta}{v} - \frac{1}{2}\frac{d}{d\theta}\frac{p\sin\theta}{v}$ reduces to

$$\sin\theta\cos\theta - \frac{1}{2}\frac{d}{d\theta}\sin^2\theta, \text{ that is to zero;}$$

and similarly for BD it also reduces to zero.

For CD this expression reduces to

$$\sin^2\theta + \frac{1}{2}\frac{d}{d\theta}\cos\theta\sin\theta, \text{ that is to } \frac{1}{2}.$$

The term outside the integral sign, namely, $-(\delta r)^2 \frac{p\sin\theta}{2v}$, will not vanish at C and at D, for $\frac{p\sin\theta}{v}$ has two values at C and D by reason of the discontinuity, namely, $\frac{p}{v} = \sin\theta$ along AC, and $\frac{p}{v} = -\cos\theta$ along CD, and then $\frac{p}{v} = -\sin\theta$ along DB, that is from D towards B.

Let r_1 and θ_1 be the co-ordinates of C, and r_2 and θ_2 the co-ordinates of D. Then finally we get for the terms of the second order in δu

$$-\frac{1}{2}(\delta r_1)^2 \sin\theta_1(\sin\theta_1 + \cos\theta_1) - \frac{1}{2}(\delta r_2)^2 \sin\theta_2(\sin\theta_2 - \cos\theta_2)$$

$$-\frac{1}{2}\int_{\theta_1}^{\theta_2} (\delta r)^2\,d\theta$$

$$+\int \frac{(k - r\sin\theta)(p\,\delta r - r\,\delta p)^2}{v^3}\,d\theta,$$

where the last integral extends over the whole discontinuous line from A to B.

Now we cannot assert that this expression of the second order is *always* positive; but we do not require that it should be so: for by the preceding Article all we require is that this expression should be positive when δr *is supposed to vanish along CD.* In this case $\delta r_1 = 0$, $\delta r_2 = 0$, and $\int_{\theta_1}^{\theta_2} (\delta r)^2 \, d\theta = 0$; so that the expression is positive.

68. The question may be asked whence does this discontinuous solution arise? The reply is that the part CD presents itself in accordance with the general principle of Art. **17**; for along that line δy is not susceptible of either sign. The two parts AC and BD are implicitly involved in the fundamental differential equation of Art. 58, namely

$$\frac{y}{\sqrt{(1+p^2)}} = \text{constant};$$

inasmuch as p equal to infinity, combined with the constant equal to zero, may be considered as a solution of the equation.

69. The conclusion of the investigation is as follows: the problem enunciated in Art. 57 *always* admits of a certain discontinuous solution, and *sometimes* admits of a certain continuous solution. When both solutions are admissible we shall find that sometimes the discontinuous solution is the less of the two, and sometimes the continuous solution.

If the two points are very near each other and at a great distance from the axis of x it is plain that the proper catenary will give a less surface than the discontinuous solution. If the given points are so situated that the two catenaries nearly coincide, the discontinuous solution gives a less surface than the proper catenary. For let S denote the surface generated by a portion of a catenary extending from the lowest point to any point P et the tangent to the catenary at P meet the axis of x at a point T the abscissa of which is ξ. Let Σ denote the surface generated by the revolution of PT round the axis of x. Then it is known that

$$S = \Sigma + \pi c \xi,$$

ξ being considered positive or negative according as P and T are on the same side of the axis of y, which is supposed to pass through the lowest point of the catenary, or on opposite sides of it. See Moigno and Lindelöf, page 212. Hence it will follow that the entire surface generated by the revolution of an arc of a catenary round the axis of x is greater than, equal to, or less than that generated by the extreme tangents according as the intersection of these tangents is below, or on, or above the axis of x. The surface generated by the tangents is of course greater than that generated by the extreme ordinates, that is greater than the discontinuous minimum surface.

If the numerical values of the extreme ordinates are given in any case, as well as their distance apart, we can by numerical calculation find the approximate value of c if there be a possible value; then calculate the surface: and so determine whether the continuous minimum is greater or less than the discontinuous minimum.

70. We have spoken throughout of a *minimum* surface. It is sufficiently obvious however that there must be a *least* surface; and as no other solution can be found there is no doubt that the *least* surface is one of the two minimum surfaces when both these exist; and when the discontinuous minimum is the only one that exists it is the least surface.

71. The preceding problem gives a simple natural illustration of the general principles which we have laid down: see the remark III. of Art. 14 and also Art. 18.

72. An important part of the preceding investigation consists in shewing that when the two catenaries coincide the tangents at the fixed points intersect on the directrix. We may easily establish this result independently.

Let the equation to the catenary be

$$y = \frac{c}{2}\left(e^{\frac{x+n}{c}} + e^{-\frac{x+n}{c}}\right);$$

let the abscissæ of the fixed points be a and h, and the corresponding ordinates b and k: so that

$$\left.\begin{aligned} b &= \frac{c}{2}\left(e^{\frac{a+n}{c}} + e^{-\frac{a+n}{c}}\right) \\ k &= \frac{c}{2}\left(e^{\frac{h+n}{c}} + e^{-\frac{h+n}{c}}\right) \end{aligned}\right\} \quad \text{.....................(1).}$$

We require that (1) should be true also when c and n receive indefinitely small increments δc and δn. Put $\psi(x)$ for $\frac{dy}{dx}$, that is for

$$\frac{1}{2}\left(e^{\frac{x+n}{c}} - e^{-\frac{x+n}{c}}\right).$$

Then from (1)

$$b\delta c - (a+n)\,\psi(a)\,\delta c - c\psi(a)\,\delta n = 0,$$

$$k\delta c - (h+n)\,\psi(h)\,\delta c - c\psi(h)\,\delta n = 0\,;$$

therefore $$\frac{b-(a+n)\,\psi(a)}{k-(h+n)\,\psi(h)} = \frac{\psi(a)}{\psi(h)}\,;$$

therefore $$\psi(h)\,\{a\psi(a) - b\} = \psi(a)\,\{h\psi(h) - k\} \text{...........(2).}$$

But the equation to the tangent to the catenary at the point (x, y) is

$$y_1 - y = \psi(x)\,(x_1 - x),$$

where x_1 and y_1 are the variable coordinates. Therefore at the intersection of the extreme tangents we have

$$\frac{y_1 - b + a\psi(a)}{y_1 - k + h\psi(h)} = \frac{\psi(a)}{\psi(h)}.$$

In order then that y_1 may be zero it is necessary and sufficient that

$$\psi(h)\,\{a\psi(a) - b\} = \psi(a)\,\{h\psi(h) - k\}\,;$$

and this agrees with (2). Thus the required result is established.

CHAPTER V.

MAXIMUM SOLID OF REVOLUTION.

73. To determine a solid of revolution the surface of which is given, so that it may cut the axis of revolution at given points and have a maximum volume.

This problem has given rise to some discussion and controversy, as will be seen by consulting the volumes of the *Philosophical Magazine* for 1866. I shall repeat with brevity what has already been established with respect to the problem, and then proceed to additional investigations.

Adopting the usual notation we have to make $\pi \int y^2 dx$ a maximum while $2\pi \int y \sqrt{(1+p^2)}\, dx$ is given; the limiting values of the variables being fixed. Thus by the usual method we require the maximum of

$$\int \{y^2 + 2ay \sqrt{(1+p^2)}\}\, dx,$$

where a is a constant at present undetermined. Denote the integral by u; then to the first order

$$\delta u = \frac{2ayp\delta y}{\sqrt{(1+p^2)}} + \int M \delta y\, dx,$$

where M stands for

$$2y + 2a \sqrt{(1+p^2)} - 2a \frac{d}{dx} \frac{yp}{\sqrt{(1+p^2)}}.$$

By the known principles of the subject we put

$$M = 0,$$

and this leads in the usual way to

$$\frac{2ay}{\sqrt{(1+p^2)}} = b - y^2.$$

Since the generating curve is to meet the axis of x we have $y = 0$ at certain points; hence $b = 0$, and the equation just obtained becomes

$$\left\{\frac{2a}{\sqrt{(1+p^2)}} + y\right\} y = 0.$$

Thus we appear to have either $\frac{2a}{\sqrt{(1+p^2)}} + y = 0$ or $y = 0$.

If we take $\frac{2a}{\sqrt{(1+p^2)}} + y = 0$, we obtain

$$p^2 = \frac{4a^2 - y^2}{y^2};$$

therefore
$$\frac{dx}{dy} = \frac{y}{\sqrt{(4a^2 - y^2)}}.$$

This gives us a circle of radius $2a$ having its centre on the axis of x. But if a circle has its centre on the axis of x and passes through two fixed points its radius is determined; and so the corresponding surface of the sphere cannot have a given value. Thus we have not a satisfactory solution of the problem.

This led to the suggestion by the Astronomer Royal that the solution of the problem is to be obtained by combining the two results; and taking $\frac{2a}{\sqrt{(1+p^2)}} + y = 0$ for part of the required line, and $y = 0$ for part of it. This gives for the solution a sphere which is connected by a straight line, namely part of the axis of revolution, with the fixed points. For facility of conception we may consider this straight line as an infinitesimally slender cylinder.

74. But on examining the proposed solution we find that the supposition $y=0$ does not make M vanish; and thus at first sight the proposed solution appears unsatisfactory.

Nevertheless there is no doubt that we have the true solution here; the apparent difficulty is removed by the principle of Art. 18. For corresponding to $y=0$ the value of δy is essentially *positive;* hence we are not compelled to have $M=0$: it will be sufficient that M be negative. Now when $y=0$ we have $M=2a$, and $2a$ is necessarily negative, as we see from the equation

$$\frac{2a}{\sqrt{(1+p^2)}}+y=0,$$

which holds when y is not zero. Thus in fact instead of having $\delta u=0$ so far as the first order of small quantities, we have δu a negative quantity of the first order: and therefore a maximum is ensured.

[This remark, which is essential to render the proposed solution admissible, was supplied by the present writer; it was the first introduction of the important principle of Article 18 into the Calculus of Variations.]

75. Thus as long as we consider a curve which is adjacent to the discontinuous line but which differs from it through the whole extent, we are sure of a maximum. But if we do not vary the part which corresponds to $y=0$, we should have to appeal to some other evidence to shew that we have secured a maximum. It is however unnecessary to investigate the terms of the second order in δu; for we may rely on a theorem which is well known, that the sphere is the body which has the greatest volume under a given surface.

76. The following then is the result:

Let A and B denote the two fixed points on the axis of revolution; then according as the given surface is greater or less than that of the sphere having AB as diameter, we have the upper diagram or the lower diagram. The generating curve must be supposed to be made up of the two rectilinear portions AC and

BE and the semicircle *CDE*. In the particular case in which the given surface is equal to that of the sphere having *AB* as diameter the rectilinear portions disappear.

The result may appear strange, at least to any person who was not familiar with the considerations brought forward in the preceding chapters of these researches: we will make a few remarks on the result.

It is certain that the sphere is the solid of greatest volume within a given surface; and thus, admitting that our solution does fulfil the prescribed conditions, we are certain that it is the greatest solid which will do so.

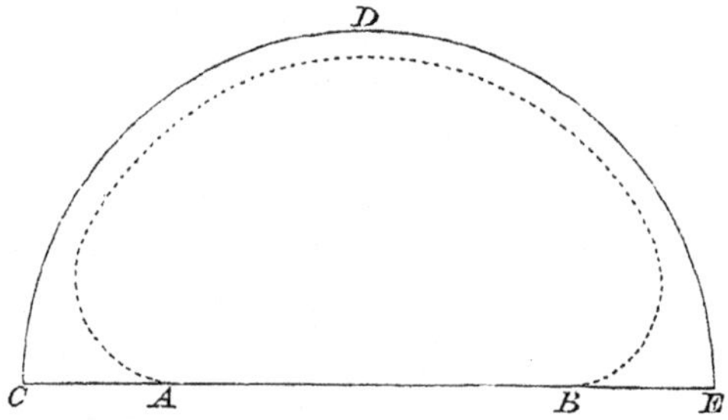

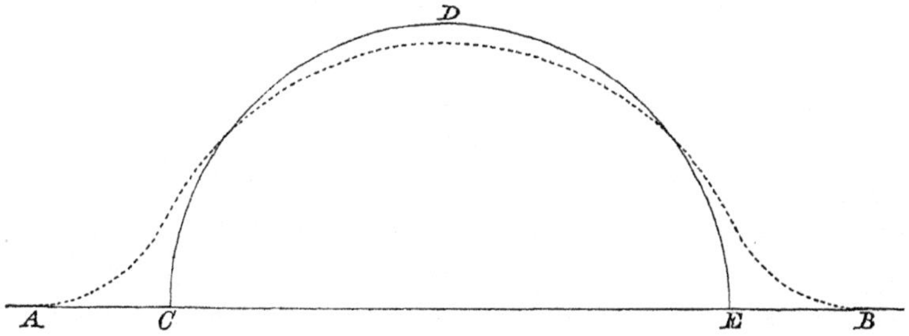

But an objector might say that he wants the greatest solid with the condition that the generating curve shall have *no abrupt change* in direction, and shall *cut* the axis instead of partly coinciding with it. I reply that nothing can be obtained which differs to an appreciable extent from the solution already given. It is obvious that we can draw a curve fulfilling the two conditions thus stated and deviating infinitesimally from the straight lines and semicircle; so that the surface and the volume will differ only to an infinitesimal extent from those of our solution: or if we make the surfaces equal the volumes will differ only infinitesimally. See the remark III. of Art. 14. A good method for drawing these curves theoretically would be to employ the propositions which serve as the foundation for the expansion of functions in terms of sines and cosines of multiple angles.

77. Abandoning then the attempt to obtain any other solution for the *greatest* solid than that which we have given, the objector may still say that he asks for a *maximum* solid among all those which have a given surface, the generating curve being constrained to cut the axis and to have no abrupt change of direction. I say that it is hopeless to seek for any such maximum distinct from what we have given. For there being no restriction introduced as to the sign of δy no one will hesitate to admit that the condition which we denote by $M=0$ *must* be satisfied; and this necessarily leads to

$$\frac{2ay}{\sqrt{(1+p^2)}} = b - y^2,$$

for the equation $M=0$ when developed is

$$2y + \frac{2a}{\sqrt{(1+p^2)}} - \frac{2ay\dfrac{dp}{dx}}{(1+p^2)^{\frac{3}{2}}} = 0,$$

that is
$$2y + \frac{2a}{\sqrt{(1+p^2)}} - \frac{2ayp\dfrac{dp}{dy}}{(1+p^2)^{\frac{3}{2}}} = 0;$$

and as this is to be true for all values of y we may integrate with respect to y: and thus we have

$$y^2 + \frac{2ay}{\sqrt{(1+p^2)}} = b.$$

Thus we cannot avoid arriving at this equation; and then we must continue the investigation as in Arts. 73 and 74.

78. We have assumed in Art. 76 that when the given surface is greater than that of a sphere having AB as diameter, the solid may stretch *beyond* the straight lines at right angles to the axis at A and B. If however the solid is restricted to lie between these straight lines, the solution is that given in Todhunter's *History of the Calculus of Variations*, page 410.

79. Although the problem enunciated in Art. 73 does not admit of solution except in the way we have explained, yet conditions may be introduced which modify the problem, and so lead to other solutions. For example, let us impose the condition that the generating curve shall never be convex to the axis of revolution, in addition to the former conditions of cutting the axis at two given points, and having no abrupt change of direction. Simple as the problem still is in enunciation it does not very obviously appear what the solution will be; and the remarks now about to be made will be confirmed or corrected if readers of the present researches will investigate the problem for themselves before consulting the solution which we shall now propose.

80. It will be convenient to change the enunciation of the problem to the following, which is of course substantially equivalent. To determine a solid of revolution of minimum surface, the volume being given; supposing that the generating curve cuts the axis at two given points, that it has no abrupt change of direction, and that it is never convex to the axis of revolution.

81. If the given volume is that of a sphere on the intercepted portion of the axis as diameter the required surface is that of this sphere; if the given volume is greater than that of this sphere the solution is that given in Todhunter's *History of the Calculus of Variations*, page 410. We have then to consider only the case in which the given volume is less than that of a sphere on the intercepted portion of the axis as diameter.

82. By the Calculus of Variations every kind of boundary of the generating figure is excluded except straight lines, and curves which satisfy the differential equation $M=0$. At first sight it

might appear that every line is excluded which does not satisfy the differential equation $M=0$: but on consideration we shall see that *straight lines are not excluded*, because it may be possible that for straight lines δy is not susceptible of either sign by reason of the condition which forbids convexity towards the axis.

Moreover the curves which satisfy the differential equation $M=0$ do not cut the axis; excluding the particular case of a semicircle which is here inadmissible: hence we are driven to the conclusion that the portion of the required boundary in the vicinity of the axis must be rectilinear.

83. I propose the following for the solution of the problem:

Let A and B be the fixed points on the axis; let AP and BQ

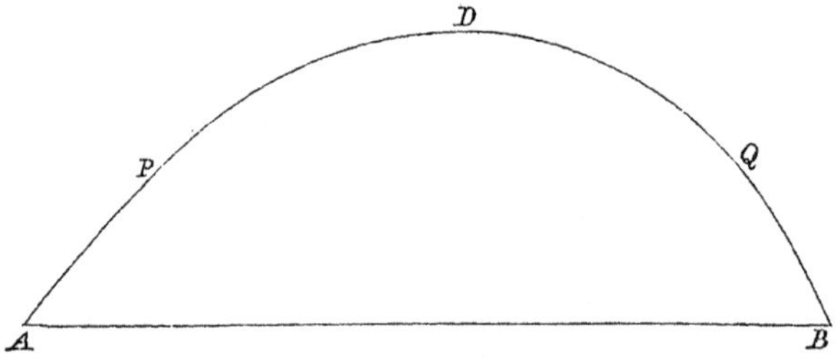

be equal straight lines equally inclined to the axis which touch the curve PQD at P and Q respectively; and let PDQ be an arc of the curve defined by the differential equation

$$\frac{2ay}{\sqrt{(1+p^2)}} = y^2 + c^2,$$

where a and c are constants. The constants must be taken so as to ensure the tangency at P and Q, and to make the volume generated by $APDQB$ have the given value. Moreover there is a certain condition to be satisfied which we shall investigate presently; this condition connects the constant a with the ordinate of P and Q, and the inclination of PA and QB to AB.

This condition is $y_1 = \frac{3a}{2}\cos\beta$, where y_1 is the ordinate of P or of Q, and β is the angle of inclination of AP or BQ to the axis AB: see Art. 85.

It will be observed that the constant c^2 corresponds to the $-b$ of Art. 73, and that the present constant a corresponds to the $-a$ of that Article.

84. We proceed to shew that in the way just stated we do obtain a minimum value of the surface. It will be seen that the differential equation for determining PDQ is a first integral of the equation $M=0$.

Let S denote the surface generated; so that

$$S = 2\pi \int y \sqrt{(1+p^2)}\, dx;$$

then to the first order,

$$\delta S = \frac{2\pi y p \delta y}{\sqrt{(1+p^2)}} + 2\pi \int \left\{ \sqrt{(1+p^2)} - \frac{d}{dx} \frac{yp}{\sqrt{(1+p^2)}} \right\} \delta y\, dx$$

$$= \frac{2\pi y p \delta y}{\sqrt{(1+p^2)}} + 2\pi \int \left\{ \frac{1}{\sqrt{(1+p^2)}} - \frac{yq}{(1+p^2)^{\frac{3}{2}}} \right\} \delta y\, dx,$$

where q stands for $\frac{dp}{dx}$. Both parts of the expression for δS are of course to be taken between limits. Now by means of the equation to PDQ we find that the coefficient of δy under the integral sign reduces to $\frac{y}{a}$ for the part PDQ of the boundary. For the rectilinear parts $q=0$, and so the coefficient of δy under the integral sign reduces to $\frac{1}{\sqrt{(1+p^2)}}$.

The term in δS which is outside the integral sign vanishes; because A and B are fixed points, and at P and Q the straight lines *touch* the curve.

Thus δS consists of $\frac{2\pi}{a}\int y \delta y\, dx$ for limits corresponding to PDQ and of $2\pi \int \frac{\delta y\, dx}{\sqrt{(1+p^2)}}$ for limits corresponding to AP and BQ; that is we may say that δS consists of $\frac{2\pi}{a}\int y\delta y\,dx$ for limits corresponding to the whole line together with $2\pi\int\left\{\frac{1}{\sqrt{(1+p^2)}}-\frac{y}{a}\right\}\delta y\,dx$ for limits corresponding to AP and BQ.

Now since the volume of the solid is given we have

$$\pi\int\{2y\,\delta y+(\delta y)^2\}\,dx=0,$$

so that to the first order of small quantities $2\pi\int y\,\delta y\,dx=0$. Thus finally δS reduces to $2\pi\int\left\{\frac{1}{\sqrt{(1+p^2)}}-\frac{y}{a}\right\}\delta y\,dx$ over limits corresponding to AP and BQ; and in order to ensure a minimum it is therefore essential that this expression should be zero or positive.

85. Let β denote the angle between AB and AP; then along AP we have $\frac{1}{\sqrt{(1+p^2)}}=\cos\beta$.

Let y_1 denote the ordinate of P. Suppose y_1 such that

$$\frac{y_1^2\cos\beta}{2}-\frac{y_1^3}{3a}=0,$$

so that

$$y_1=\frac{3a}{2}\cos\beta.$$

Then it is obvious that if we take δy proportional to y we have $\int\left(\cos\beta-\frac{y}{a}\right)\delta y\,dx=0$, for limits corresponding to AP. For if we put $\delta y=\mu y$, where μ is a constant, the value of the integral between the limits corresponding to AP, is $\mu\left(\frac{y_1^2\cos\beta}{2}-\frac{y_1^3}{3a}\right)$, that is zero. And we shall now shew that the integral will be *positive* for any other supposition respecting δy consistent with the condition that there is *no convexity*. Throughout when we speak of taking δy proportional to y, we mean that it is so along AP or BQ; for other parts of the boundary δy may be taken as we please.

For let $ALIK$ be a curve obtained from the straight line AP

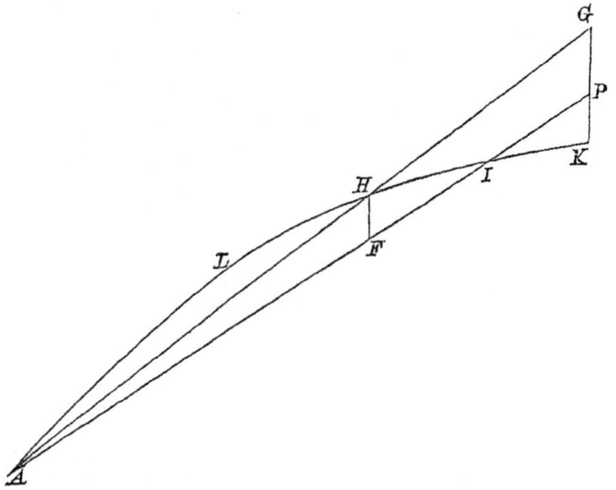

by ascribing admissible values to δy: so that the curve has no convexity towards the axis of revolution, and no abrupt change of direction.

Let $AF = \frac{2}{3} AP$. Draw FH parallel to the axis of y to meet the curve at H, and draw the straight line AHG.

Now if δy relate to the straight line AHG we have δy proportional to y; and then $\int \left(\cos\beta - \frac{y}{a}\right) \delta y \, dx$ for the limits with which we are concerned is *zero*; that is $\int \left(\frac{2y_1}{3} - y\right) \delta y \, dx$ is zero. Hence we shall find that if δy relate to the curve $ALHK$ we must have a *positive* value for $\int \left(\frac{2y_1}{3} - y\right) \delta y \, dx$.

For now there is a gain of positive elements corresponding to the area between AH and ALH; there is a diminution of negative elements in having HIF instead of $HFPG$, that is a relief from the negative elements corresponding to the area $HIPG$; and there is a gain of positive elements corresponding to the area IKP.

The preceding diagram supposes that FH has to be drawn *upwards* to meet the curve; if FH has to be drawn *downwards* we have such diagrams as the following:

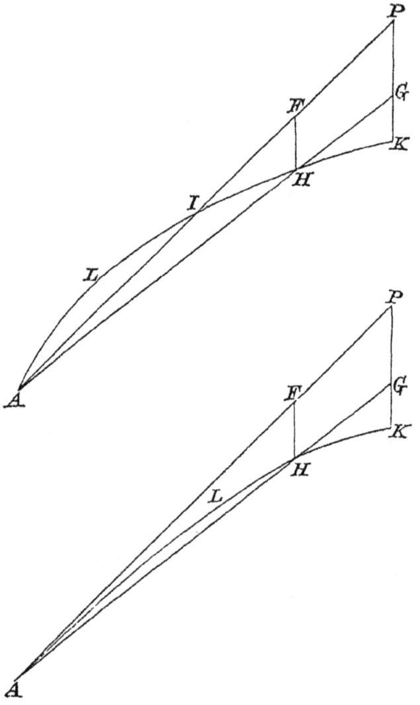

In both cases the integral $\int\left(\frac{2y_1}{3} - y\right)\delta y\,dx$ is zero when δy relates to the straight line AG; and hence we shall find that the integral is positive when δy relates to the curve AHK.

For in the upper diagram there is a gain of positive elements corresponding to the area ALI, a gain of positive elements corresponding to the area HKG, and a relief from the negative elements corresponding to the area AIH.

In the lower diagram there is a gain of positive elements corresponding to the area HKG, and a relief from the negative elements corresponding to the area ALH.

86. Thus we shew that for the line $APBQ$ we have to the first order δS always positive except in one particular case, and then δS is zero. Such a result is as much as can be obtained in most problems of the Calculus of Variations; although strictly speaking in the case in which δS is zero to the first order we ought to examine the terms of the second order to be absolutely certain of a minimum. I shall return to this point at the end of the present chapter.

Of course even when we have secured a *minimum* the result is not necessarily the *least* which is possible; but in the present case I think there can be little doubt that our result is really the *least*. There must obviously be some *least* value; and the boundary must be composed of a straight line or straight lines and arcs of the curve determined by $M=0$; and I believe that with due consideration every combination of these admissible elements, except that which we have adopted, will be excluded.

87. Let us now examine more closely the equation

$$y_1 = \frac{3a}{2} \cos \beta,$$

which we have obtained for determining the points P and Q.

It is well known that the equation

$$\frac{2ay}{\sqrt{(1+p^2)}} = y^2 + c^2$$

belongs to the curve traced out by the focus of an ellipse as the ellipse rolls on the axis of x. See a memoir by Lindelöf in the *Acta Soc. Sci. Fenn.*, Helsingfors, 1863. The major axis of the rolling ellipse is $2a$, and $c^2 = a^2(1-e^2)$, where e is the eccentricity. The curve thus generated is partly concave towards the axis of x, and partly convex. The distance from the axis of the highest point is $a(1+e)$.

We are here concerned with the concave portion. If ρ be the radius of curvature at any point of the concave portion we have

$$\frac{1}{\sqrt{(1+p^2)}}+\frac{y}{\rho}=\frac{y}{a}.$$

Thus at the points P and Q since the curve touches the straight lines we have

$$\frac{2y_1}{3a}+\frac{y_1}{\rho}=\frac{y_1}{a};$$

therefore $$\rho=3a.$$

Now the radius of curvature at the point of the curve which is most distant from the axis of x will be found to be $\frac{a(1+e)}{e}$; and at the point of inflexion it is of course infinite. Thus we require that the value $3a$ should lie between $\frac{a(1+e)}{e}$ and infinity; that is e must be greater than $\frac{1}{2}$.

We may determine the value of y_1. Substituting in the equation

$$\frac{2ay}{\sqrt{1+p^2}}=y^2+a^2(1-e^2),$$

we obtain $$\frac{4y_1^2}{3}=y_1^2+a^2(1-e^2);$$

therefore $$y_1^2=3a^2(1-e^2).$$

If $e=\frac{1}{2}$ we have $y_1=\frac{3a}{2}$, and therefore $\cos\beta=1$. Thus when $e=\frac{1}{2}$ the straight lines AP and BQ coincide with the axis of revolution, so that the volume and the surface vanish. This is consistent with what we have already found, namely that $\frac{1}{2}$ is a limiting value of e.

88. Hence the following is our process of solution. Take a curve defined by the equation $\frac{2ay}{\sqrt{(1+p^2)}} = y^2 + a^2(1-e^2)$, and draw tangents at the points for which the radius of curvature is $3a$; then we have to fulfil the following conditions: these tangents must intersect the axis of x at given points, and the solid generated by the revolution of the figure round the axis of x must have a given volume. The conditions must serve for determining the constants a and e, as well as the constant which may be conceived to arise from integrating the differential equation to the curve. But practically we shall not be obliged to pay any regard to the constant introduced by integrating, since we may suppose the origin to be at any point we please in the axis of revolution.

89. Although we have now carried the solution as far as such solutions are usually carried, yet we will continue the investigation and shew that the conditions by which a and e are to be determined can always be satisfied.

As the relation between x and y for the curve PDQ cannot be exhibited explicitly, we have to adopt an indirect method.

Let $AB = 2h$; then

$$h = y_1 \cot \beta + \int_{y_1}^{a(1+e)} \frac{dy}{p}.$$

I intend to shew from this equation, in which h is a fixed quantity, that a and e increase together.

Let $$V = \frac{\pi y_1^3 \cot \beta}{3} + \pi \int_{y_1}^{a(1+e)} \frac{y^2 dy}{p}.$$

Then $2V$ denotes the volume of the solid generated by the revolution of our proposed boundary. I intend to shew that as a and e increase V continually increases, so that we can make our volume equal to any assigned volume lying between zero and the volume of the sphere having $2h$ for diameter.

90. We have

$$y_1 = \frac{3a}{2} \cos \beta, \text{ and } \cos^2 \beta = \frac{4}{3}(1-e^2).$$

As $p=0$ when $y=a(1+e)$ we shall find it convenient to change the independent variable from y to p in our integrals.

From the equation

$$\frac{2ay}{\sqrt{(1+p^2)}}=y^2+c^2$$

we get
$$y=\frac{a}{\sqrt{(1+p^2)}}\pm\sqrt{\left\{\frac{a^2}{1+p^2}-c^2\right\}};$$

the upper sign applies to the part of the curve with which we are concerned: thus

$$y=\frac{a}{\sqrt{(1+p^2)}}+a\sqrt{\left\{\frac{1}{1+p^2}-(1-e^2)\right\}},$$

$$-\frac{dy}{dp}=\frac{ap}{(1+p^2)^{\frac{3}{2}}}\left\{1+\frac{1}{\sqrt{1-(1-e^2)(1+p^2)}}\right\}.$$

Therefore

$$h=\frac{3a}{2}\frac{\cos^2\beta}{\sin\beta}+a\int_0^{\tan\beta}\frac{1}{(1+p^2)^{\frac{3}{2}}}\left\{1+\frac{1}{\sqrt{e^2(1+p^2)-p^2}}\right\}dp.$$

Let L denote the expression under the integral sign; and let L_1 be the value of L when $p=\tan\beta$. Let δa and δe denote simultaneous indefinitely small changes in a and e consistent with the relation just expressed, in which h is given. We obtain then

$$0=\frac{h}{a}\delta a+\left(\frac{3a}{2}\frac{d}{d\beta}\frac{\cos^2\beta}{\sin\beta}+aL_1\frac{d}{d\beta}\tan\beta\right)\frac{d\beta}{de}\delta e$$
$$+a\delta e\int_0^{\tan\beta}\frac{dL}{de}dp.$$

I shall shew that the coefficient of δe in this equation is *negative.*

It is obvious that $\frac{dL}{de}$ is negative.

$$\frac{d}{d\beta}\frac{\cos^2\beta}{\sin\beta}=-\left(2\cos\beta+\frac{\cos^3\beta}{\sin^2\beta}\right).$$

It will be found that $L_1=3\cos^3\beta$.

Thus the term involving $\frac{d\beta}{de}\delta e$ is

$$\left\{-\frac{3a}{2}\left(2\cos\beta+\frac{\cos^3\beta}{\sin^2\beta}\right)+3a\cos\beta\right\}\frac{d\beta}{de}\delta e,$$

that is,
$$-\frac{3a}{2}\frac{\cos^3\beta}{\sin^2\beta}\frac{d\beta}{de}\delta e.$$

Hence finally

$$\frac{h}{a}\delta a=\left\{\frac{3a}{2}\frac{\cos^3\beta}{\sin^2\beta}\frac{d\beta}{de}+\lambda\right\}\delta e,$$

where λ stands for the positive quantity

$$-a\int_0^{\tan\beta}\frac{dL}{de}dp.$$

And $\frac{d\beta}{de}$ is positive for

$$\sin\beta\cos\beta\frac{d\beta}{de}=\frac{4e}{3};$$

so that δa and δe have the same sign.

91. Now we have

$$V=\frac{9\pi a^3}{8}\frac{\cos^4\beta}{\sin\beta}+\pi a^3\int_0^{\tan\beta}\frac{\{1+\sqrt{e^2(1+p^2)-p^2}\}^3}{(1+p^2)^{\frac{5}{2}}\sqrt{e^2(1+p^2)-p^2}}dp.$$

Let H denote the expression under the integral sign, and let H_1 be the value of H when $p=\tan\beta$. Then supposing a and e to receive infinitesimal changes, and denoting by δV the consequent change in V, we have

$$\delta V=\frac{3V}{a}\delta a+\left(\frac{9\pi a^3}{8}\frac{d}{d\beta}\frac{\cos^4\beta}{\sin\beta}+\pi a^3H_1\frac{d\tan\beta}{d\beta}\right)\frac{d\beta}{de}\delta e$$
$$+\pi a^3\delta e\int_0^{\tan\beta}\frac{dH}{de}dp.$$

Now
$$\frac{dH}{de}=\frac{e\{1+\sqrt{e^2(1+p^2)-p^2}\}^2\{2\sqrt{e^2(1+p^2)-p^2}-1\}}{(1+p^2)^{\frac{3}{2}}\{e^2(1+p^2)-p^2\}^{\frac{3}{2}}}.$$

The sign of this depends on the sign of the factor

$$2\sqrt{e^2(1+p^2)-p^2}-1;$$

this expression vanishes when $p = \tan\beta$, and increases as p diminishes, so that it is always *positive.*

$$\frac{d}{d\beta}\frac{\cos^4\beta}{\sin\beta} = -\left(4\cos^3\beta + \frac{\cos^5\beta}{\sin^2\beta}\right).$$

It will be found that $H_1 = \frac{27}{4}\cos^5\beta$.

Thus the term involving $\frac{d\beta}{de}\delta e$ is

$$\frac{9\pi a^3}{8}\frac{3\sin^2\beta - 1}{\sin^2\beta}\cos^3\beta\frac{d\beta}{de}\delta e.$$

Hence $\delta V = \frac{3V}{a}\delta a + \left(\frac{9\pi a^3}{8}\frac{3\sin^2\beta-1}{\sin^2\beta}\cos^3\beta\frac{d\beta}{de} + \eta\right)\delta e,$

where η stands for the positive quantity

$$\pi a^3\int_0^{\tan\beta}\frac{dH}{de}dp.$$

In the expression for δV let us put for δe its value in terms of δa, and also the value of $\frac{d\beta}{de}$, which were both given in the preceding Article; thus we get

$$\delta V = \left\{3V + \frac{3\pi h a^3 e(3\sin^2\beta - 1) + 2h\eta\sin\beta\tan^2\beta}{4ae + 2\lambda\sin\beta\tan^2\beta}\right\}\frac{\delta a}{a}.$$

92. I shall now shew that V continually increases with a; that is that $\frac{dV}{da}$ is always positive. Of course we need only consider the case in which $\sin^2\beta$ is *less* than $\frac{1}{3}$; for when $\sin^2\beta$ is greater than $\frac{1}{3}$ we have $\frac{dV}{da}$ necessarily positive.

We may obviously put the value of $\frac{dV}{da}$ thus:

$$\frac{dV}{da} = \frac{1}{a}\{3V + ha^2\phi(e)\},$$

where $\phi(e)$ involves only e, and β, which is a function of e.

[And $$\phi(e) = \frac{3\pi(3\sin^2\beta - 1)\cos^2\beta + \dfrac{2\eta\sin^3\beta}{ea^3}}{4\cos^2\beta + \dfrac{2\lambda\sin^3\beta}{ea}}$$

$$= \psi(e) - \frac{3\pi}{4},$$

where $$\psi(e) = \frac{9\pi\cos^2\beta + \dfrac{3\pi\lambda}{2ea}\sin\beta + \dfrac{2\eta\sin\beta}{ea^3}}{4\cos^2\beta + \dfrac{2\lambda\sin^3\beta}{ea}}\sin^2\beta.$$

Therefore, substituting the value of V from Art. 91, we have

$$\frac{dV}{da} = a\left\{\frac{27\pi a}{8}\frac{\cos^4\beta}{\sin\beta} + 3\pi a\int_0^{\tan\beta} H dp + h\psi(e) - \frac{3\pi h}{4}\right\};$$

then, substituting the value of h from Art. 90, we obtain

$$\frac{dV}{da} = a^2\left\{\frac{9\pi}{8}(3\cos^2\beta - 1)\frac{\cos^2\beta}{\sin\beta} + h\psi(e) + \frac{3\pi}{4}\int_0^{\tan\beta}(4H - L)\,dp\right\}.$$

Now $3\cos^2\beta - 1 = 2 - 3\sin^2\beta$; and this is positive since $\sin^2\beta$ is less than $\frac{1}{3}$.

Also $4H - L$ is positive; for the sign of this is the sign of

$$4\{1 + \sqrt{e^2(1+p^2) - p^2}\}^2 - (1+p^2);$$

and between the limits we have to consider $1 + p^2$ is less than $\frac{1}{\cos^2\beta}$, that is less than $\frac{3}{2}$.

Hence $\frac{dV}{da}$ is positive.

The above demonstration is substituted for that which was originally offered. In the original demonstration a property of $a^2\phi(e)$ was employed, and as this property may be of use it will be given in the next Article. The reader who wishes to avoid an interruption to the reasoning may pass on to Art. 94.]

93. We have $$a^2\phi(e) = \frac{\dfrac{3\pi(3\sin^2\beta - 1)\cos^2\beta}{\sin^3\beta} + \dfrac{2\eta}{ea^3}}{\dfrac{4\cos^2\beta}{a^2\sin^3\beta} + \dfrac{2\lambda}{ea^3}}.$$

We shall find that as e increases the denominator continually diminishes, and the numerator continually increases until $\sin^2\beta$ is greater than $\frac{1}{3}$.

First consider the denominator.

$$\frac{\lambda}{ea^3} = \frac{1}{a^2}\int_0^{\tan\beta} \frac{1}{\sqrt{(1+p^2)}} \times \frac{1}{\{e^2(1+p^2)-p^2\}^{\frac{3}{2}}}\,dp.$$

Put R for $\dfrac{4\cos^2\beta}{a^2\sin^3\beta} + \dfrac{2\lambda}{ea^3}$.

Then $\dfrac{dR}{da}$ consists of various terms every one of which is necessarily negative, except that which arises from the variation of the upper limit in the integral involved in the value of λ; the term in $\dfrac{dR}{da}$ which thus arises is

$$\frac{2}{a^2}\cos\beta \cdot \frac{1}{\cos^2\beta} \cdot \frac{1}{\left(\frac{1}{4}\right)^{\frac{3}{2}}} \frac{d\beta}{de}\frac{de}{da},$$

that is

$$\frac{16}{a^2\cos\beta}\frac{d\beta}{de}\frac{de}{da}.$$

But $\dfrac{dR}{da}$ besides other negative terms has

$$\frac{4}{a^2}\frac{d}{d\beta}\left(\frac{\cos^2\beta}{\sin^3\beta}\right)\frac{d\beta}{de}\frac{de}{da},$$

that is,

$$-\frac{8\cos\beta\sin^2\beta + 12\cos^3\beta}{a^2\sin^4\beta}\frac{d\beta}{de}\frac{de}{da};$$

and this negative term more than counterbalances the positive term just given; for $(8\cos\beta\sin^2\beta + 12\cos^3\beta)\cos\beta$ is greater than $16\sin^4\beta$, provided $\sin^2\beta$ is less than $\frac{1}{3}$. To shew this we must compare $\cos^2\beta(2+\cos^2\beta)$ with $4\sin^4\beta$; we shall find that $\cos^2\beta(2+\cos^2\beta) - 4\sin^4\beta = 3 - 4\sin^2\beta - 3\sin^4\beta$, and this is positive since $\sin^2\beta$ is less than $\frac{1}{3}$.

Next consider the numerator.

$$\frac{\eta}{ea^3} = \pi \int_0^{\tan\beta} \frac{dp}{(1+p^2)^{\frac{3}{2}}} \frac{(1+x)^2(2x-1)}{x^3},$$

where x is put for $\sqrt{e^2(1+p^2)-p^2}$.

The integral certainly increases with e if $\frac{(1+x)^2(2x-1)}{x^3}$ increases with x.

Now $$\frac{(1+x)^2(2x-1)}{x^3} = 2 + \frac{3}{x} - \frac{1}{x^3};$$

the differential coefficient of this with respect to x is $\frac{3(1-x^2)}{x^4}$, which is positive; for $x^2 = e^2 - (1-e^2)p^2$ and is therefore less than unity.

And $$\frac{(3\sin^2\beta - 1)\cos^2\beta}{\sin^3\beta} = \frac{(3\sin^2\beta - 1)(1-\sin^2\beta)}{\sin^3\beta}$$

$$= -3\sin\beta + \frac{4}{\sin\beta} - \frac{1}{\sin^3\beta};$$

the differential coefficient of this with respect to β is

$$\frac{\cos\beta(3 - 4\sin^2\beta - 3\sin^4\beta)}{\sin^4\beta}:$$

now this is positive when $\beta = 0$, and does not change sign so long as $\sin^2\beta$ is less than $\frac{1}{3}$, in fact not until $\sin^2\beta$ is greater than $\frac{1}{2}$. Thus denoting the numerator of $a^2\phi(e)$ by T we have shewn that $\frac{dT}{da}$ is positive.

94. Thus we have shewn that $\frac{dV}{da}$ is always positive. If this had not been established we should have been in the following position; instead of knowing that V continually *increases* from 0 to $\frac{4\pi h^3}{3}$, we could only assert that V *changes* from 0 to $\frac{4\pi h^3}{3}$. Thus

the same value of V might result from different values of a; and in consequence corresponding to a given volume there might be more than one minimum surface. In general we should expect that these surfaces would differ in area, so that although both or all were *minima* surfaces every one could not be the *least*. This result would not be in any way repugnant to the principles of the Calculus of Variations. However by our demonstration that $\frac{dV}{da}$ is always positive we see that our solution is unique.

95. As we have already stated the extreme case of our solution is that in which we have for the generating curve a semicircle on the given part of the axis as diameter; see Arts. 81 and 89. When the given volume is greater than corresponds to this case the required solution is that given in Todhunter's *History of the Calculus of Variations*, page 410. We shall however now add some remarks similar to those in Arts. 89...94, in order to shew that the conditions relating to the constants can be satisfied.

96. The boundary which we are now about to consider is

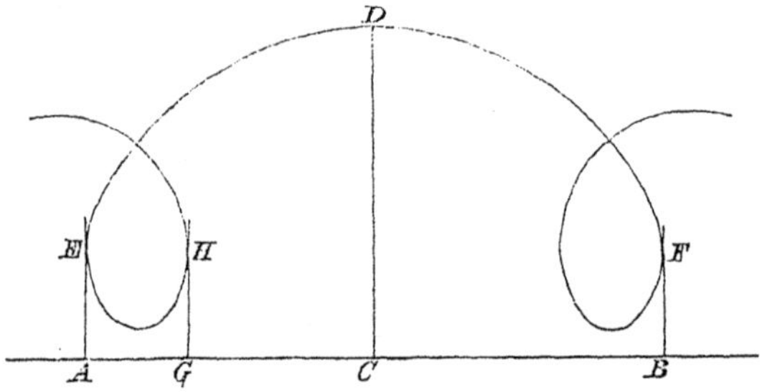

composed of two equal straight lines AE and BF, and the curve EDF. The straight lines are at right angles to the axis at the given points, and they touch the curve at the points E and F. The curve is determined by the equation

$$\frac{2ay}{\sqrt{(1+p^2)}} = y^2 - c^2 \text{..........................} (1),$$

where a and c are constants.

The curve is traced out by the focus of an hyperbola as the hyperbola rolls on the axis of x: see the memoir by Lindelöf, already cited in Art. 87. The curve consists of an endless repetition of portions like that in the diagram. HG is the tangent at H, and is equal and parallel to AE.

And $c^2 = a^2(e^2 - 1)$, where e is the eccentricity of the hyperbola.

Moreover the following relations hold:

$$CD = a(1+e), \quad AE = c = a\sqrt{(e^2-1)},$$

$$AB = 2a + 2a\int_0^{\sin^{-1}\frac{1}{e}} \sqrt{(1 - e^2\sin^2\theta)}\, d\theta,$$

$$AG = 2a - 2a\int_0^{\sin^{-1}\frac{1}{e}} \sqrt{(1 - e^2\sin^2\theta)}\, d\theta.$$

97. We take AB to be fixed, and we shall shew that a and e increase together; and that as they increase we obtain a set of boundaries like $AEDFB$ the curved part of each of which is outside that of its predecessor.

Let $AB = 2h$. When $c = 0$ the boundary consists of the semicircle

$$\frac{hy}{\sqrt{(1+p^2)}} = y^2 \quad \text{..........................} \quad (2).$$

From the general formulæ of Art. 96 we have $AB + AG = 4a$; and as $AB = 2h$ we have $2h$ less than $4a$, and therefore h less than $2a$.

The curve (1) begins by being *above* the curve (2) at the points in the neighbourhood of A and B. Suppose if possible that (1) and (2) could intersect. At the points of intersection nearest to the axis of x let p_1 be the value of p for the curve (1), and p_2 the value of p for the curve (2), that is for the semicircle. Then p_2 is greater than p_1; and as the curves intersect we have at the common point

$$\frac{hy}{\sqrt{(1+p_2^2)}} = \frac{2ay}{\sqrt{(1+p_1^2)}} + c^2.$$

But this is impossible; for by what has been shewn $\frac{h}{\sqrt{(1+p_2^2)}}$ is less than $\frac{2a}{\sqrt{(1+p_1^2)}}$.

Thus any curve given by (1) with the relations of Art. 96 is outside the curve given by (2).

98. We have thus compared the curve (1) with the curve (2); we now proceed to compare together two curves determined by (1) with different values of the constants a and e.

Since

$$h = a + a\int_0^{\sin^{-1}\frac{1}{e}} \sqrt{(1 - e^2 \sin^2\theta)}\, d\theta,$$

we have

$$0 = \frac{h}{a}\,\delta a - ae\,\delta e \int_0^{\sin^{-1}\frac{1}{e}} \frac{\sin^2\theta\, d\theta}{\sqrt{(1 - e^2\sin^2\theta)}};$$

this shews that δa and δe have the same sign, so that a and e increase together. Therefore of course c also increases with a, for

$$c^2 = a^2(e^2 - 1).$$

Now take the two curves

$$\frac{2a_1 y}{\sqrt{(1+p^2)}} = y^2 - c_1^2,$$

and

$$\frac{2a_2 y}{\sqrt{(1+p^2)}} = y^2 - c_2^2;$$

suppose a_1 greater than a_2, and therefore c_1 greater than c_2. Then the former curve is above the latter at the points in the neighbourhood of E and F; and so by the method of argument already used in Art. 97, the former curve is entirely above the latter for the portion with which we are concerned.

Thus we see that as a increases the volume generated by the revolution of our boundary continually increases; and so a boundary exists for any assigned volume which is greater than that of a sphere having AB as diameter.

99. We may establish in another way the result obtained in Arts. 91...93, namely that when the volume is less than that of the limiting sphere the volume continually increases with a.

We have
$$\frac{2ay}{\sqrt{(1+p^2)}} = y^2 + c^2,$$
where
$$c^2 = a^2(1-e^2).$$

Now if c decreased as a increased we should see by the method of argument used in Art. 97 that the two curves

$$\frac{2a_1 y}{\sqrt{(1+p^2)}} = y^2 + c_1^2 \text{ and } \frac{2a_2 y}{\sqrt{(1+p^2)}} = y^2 + c_2^2$$

could not intersect during the range of values with which we are concerned, that is the range between the points P and Q of the diagram of Art. 83.

For suppose a_1 greater than a_2; then e_1 is greater than e_2; thus $\frac{4}{3}(1-e_1^2)$ is less than $\frac{4}{3}(1-e_2^2)$, and so the angle of inclination of AP to AB is greater for the curve corresponding to a_1 and e_1 than for the curve corresponding to a_2 and e_2. Hence at the points in the neighbourhood of P and Q the former curve is *above* the latter curve.

Then if the two curves could intersect, we should have at a common point, as in Art. 97,

$$\frac{2a_1 y}{\sqrt{(1+p_1^2)}} - c_1^2 = \frac{2a_2 y}{\sqrt{(1+p_2^2)}} - c_2^2;$$

but, as p_1 is less than p_2 at the lowest point of intersection, if c_1 is less than c_2, the left-hand member of this supposed equation would be greater than the right-hand member. Thus the curves could not intersect.

But $c\frac{dc}{da} = a(1-e^2) - ea^2\frac{de}{da}$, so that $\frac{dc}{da}$ is negative when $e=1$. Therefore as we approach the limiting case of the semicircle for the generating curve, the curve for which a and e have specific values is necessarily inside the curve for which a and e have infinitesimally greater values.

If it be possible let a_1 and e_1 denote such simultaneous values

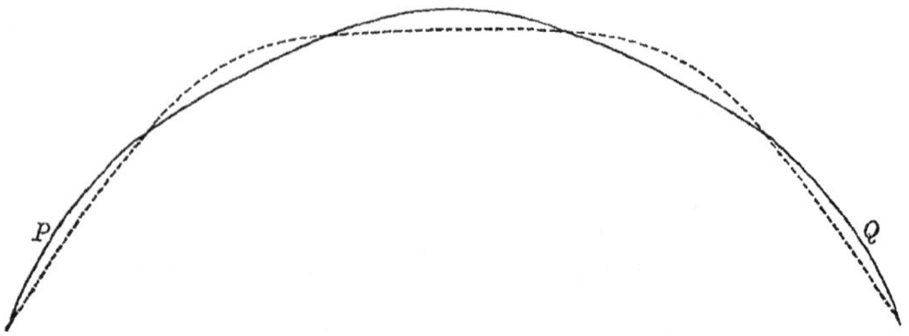

of a and e that the corresponding curve is entirely inside the curve for which the simultaneous values are $a_1 + \delta a_1$ and $e_1 + \delta e_1$, while it is intersected by the curve for which the simultaneous values are $a_1 - \delta a_1$ and $e_1 - \delta e_1$. The last curve is *below* the curve corresponding to a_1 and e_1 both at the vertex and at the points in the neighbourhood of P and Q, where the curve touches the straight lines: hence the supposed intersections must be at points intermediate between P and Q and the highest points. But this is impossible, for when δa_1 and δe_1 are small enough the curve corresponding to a_1 and e_1 would intersect the curve corresponding to $a_1 + \delta a_1$ and $e_1 + \delta e_1$ at the same point: but by supposition the last two curves do *not* intersect.

100. It is obvious that the whole problem illustrates our principle that discontinuity arises from a condition imposed. In Arts. 73...79 there is a condition implicitly imposed, namely that the generating curve is to be entirely on the positive side of the axis. In Arts. 80...99 we have the condition of concavity explicitly introduced.

101. I proceed now as stated in Art. 86 to consider whether S is really a minimum.

If we ascribe a variation to the rectilinear parts of the boundary we are certain that δS is a positive quantity of the first order, except in one particular case in which δS to the first order is zero. We will now apply Jacobi's method.

102. Let $u=\int\left\{2y\sqrt{(1+p^2)}-\frac{y^2}{a}\right\}dx$, the integral being taken between fixed limits; and suppose that we transform by Jacobi's method the term of the second order in δu.

In order that there may be a maximum or minimum value of u we must have

$$\frac{2ay}{\sqrt{(1+p^2)}}=y^2+c_1 \quad\text{.........................(1)},$$

then

$$x=\int\frac{dy}{p}+c_2 \quad\text{.........................(2)}.$$

Here c_1 and c_2 are arbitrary constants.

Suppose p found from (1) and substituted in (2); then (2) becomes the relation between x, y, c_1 and c_2, which is equivalent to the equation $y=f(x, c_1, c_2)$ of Art. 25. Hence to find $\frac{df}{dc_1}$ and $\frac{df}{dc_2}$, we have from (2)

$$0=-\int\frac{1}{p^2}\frac{dp}{dc_1}dy+\frac{1}{p}\frac{df}{dc_1},$$

$$0=\frac{1}{p}\frac{df}{dc_2}+1.$$

Thus

$$\frac{df}{dc_1}=-p\int\frac{1}{p^2}\frac{dp}{dc_1}dy,$$

$$\frac{df}{dc_2}=-p.$$

From (1) we have

$$-\frac{2ayp}{(1+p^2)^{\frac{3}{2}}}\frac{dp}{dc_1}=1,$$

so that

$$\frac{df}{dc_1}=\frac{p}{2a}\int\frac{(1+p^2)^{\frac{3}{2}}}{yp^3}dy.$$

Hence the value of z required in Art. 25 is theoretically found. Although we cannot express $\frac{df}{dc_1}$ explicitly in finite terms yet the form given for it will suffice for our purpose.

103. Now if we extend the process of Art. 84 so as to include terms of the second order, we shall have

$$\delta S = 2\pi \int \left\{ \frac{1}{\sqrt{(1+p^2)}} - \frac{y}{a} \right\} \delta y \, dx$$

$$+ \pi \int \left\{ \frac{y(\delta p)^2}{(1+p^2)^{\frac{3}{2}}} + \frac{2p\delta p \delta y}{(1+p^2)^{\frac{1}{2}}} - \frac{(\delta y)^2}{a} \right\} dx,$$

where the former integral extends over the rectilinear parts of the solution, and the latter integral over the whole solution. We are concerned now only with the latter integral. By Art. 23 this can be transformed to

$$\frac{\pi}{2} \int \frac{y}{(1+p^2)^{\frac{3}{2}}} \left(\delta p - \frac{z'}{z} \delta y \right)^2 dx.$$

Now, in the manner of Art. 24, we have

$$z = C_2 \frac{df}{dc_2} (1 + mv),$$

where m stands for $\frac{C_1}{C_2}$ and v for $\frac{\frac{df}{dc_1}}{\frac{df}{dc_2}}$; and we are sure of a minimum provided that v does not range over all values between positive infinity and negative infinity.

Here $$v = -\frac{1}{2a} \int \frac{(1+p^2)^{\frac{3}{2}}}{yp^3} dy;$$

in this expression put for y its value in terms of p as in Art. 90; thus we obtain

$$v = \frac{1}{2a} \int \frac{\sqrt{(1+p^2)} \, dp}{p^2 \sqrt{e^2 - (1-e^2) p^2}}.$$

If p is very small we have

$$v = \frac{1}{2ae}\int \frac{1}{p^2}\left\{1 + \frac{p^2}{2e^2} + \dots\right\} dp,$$

that is

$$v = \frac{1}{2ae}\left\{-\frac{1}{p} + \frac{p}{2e^2} + \dots\right\}.$$

Thus if p is indefinitely small and positive, v is negative and numerically indefinitely large.

As $\frac{dv}{dp}$ is positive v increases with p; thus two cases exist: either v becomes positive before p arrives at the greatest admissible value, which with the notation of Art. 90 is $\tan\beta$, or v remains negative up to the limit. Moreover v changes sign with p. Hence in the second case v does *not* range over every value between positive infinity and negative infinity; and so we are certain of a minimum. In the former case however v *does* range over every value between positive infinity and negative infinity; and so it might appear that there is really not a minimum. The inference however would be erroneous; all we can say is that Jacobi's transformation becomes inapplicable. Jacobi's method is quite satisfactory for problems of *absolute* maxima and minima values, but not for problems of *relative* maxima and minima values. In the present problem we have the important condition that the volume is to be constant; this imposes certain restrictions on δy: for instance δy cannot be positive throughout or negative throughout.

104. Let us take a special case. Suppose $e = 1$: then

$$v = \frac{1}{2a}\int \frac{\sqrt{(1+p^2)}\, dp}{p^2}.$$

Put $\tan\theta$ for p; thus

$$v = \frac{1}{2a}\int \frac{d\theta}{\cos\theta \sin^2\theta}.$$

It will be found that

$$v = \frac{1}{2a}\left\{-\frac{1}{\sin\theta} + \frac{1}{2}\log\frac{1+\sin\theta}{1-\sin\theta}\right\}.$$

In this case $\beta = \frac{\pi}{2}$; and so v ranges from positive infinity to negative infinity while θ ranges between 0 and β, and again while θ ranges between 0 and $-\beta$. Thus if Jacobi's method were applicable we should infer that there is not a minimum; but in this case our solution becomes a sphere, which we know gives the least surface with a given volume.

105. The general value of v becomes by putting $\tan\theta$ for p

$$v = \frac{1}{2a}\int \frac{d\theta}{\sin^2\theta\sqrt{(e^2 - \sin^2\theta)}};$$

thus $$v = \frac{1}{2ae}\int \frac{d\theta}{\sin^2\theta} + \frac{1}{2ae}\int \frac{d\theta}{\sin^2\theta}\left\{\left(1 - \frac{\sin^2\theta}{e^2}\right)^{-\frac{1}{2}} - 1\right\};$$

therefore the sign of v is the same as that of

$$-\cot\theta + \int \frac{d\theta}{\sin^2\theta}\left\{\left(1 - \frac{\sin^2\theta}{e^2}\right)^{-\frac{1}{2}} - 1\right\}.$$

The last integral is always finite; we may suppose it taken between the limits 0 and θ, so that it vanishes with θ, and increases with θ. Thus if θ be positive we see that v is $-\infty$ when $\theta = 0$; at the limit β we have $\cot\theta = \cot\beta = \sqrt{\frac{4 - 4e^2}{4e^2 - 1}}$, and so the sign of v may be positive or negative at this limit according to the value of e.

106. Suppose the value of e to be such that $v = 0$ when $p = \tan\beta$; then v just ranges between positive infinity and negative infinity. Thus if Jacobi's method were applicable we should infer that there is not a minimum; we shall shew however that there is certainly a minimum.

Whatever be the system of values of δy we must have δy vanishing at some point or points besides the extreme points; otherwise δy would be of the same sign throughout, and this is impossible. Suppose that δy vanishes for the point of the curve at which $p = \varpi$; take m such that $1 + mv$ vanishes at this point. Thus z vanishes when $p = \varpi$, and does not vanish at any other point. Hence this value of z satisfies the conditions involved in the investigation of Jacobi's theorem; and thus the term of the second order in δS takes the essentially positive form given in Art. 103.

107. We may observe that our solution is certainly a minimum when compared with all such solids as have the same vertex as itself.

For with this condition we always have δy zero when p is zero. Hence instead of taking the general value of z we may take the particular value $C_3\, p$, where C_3 is any constant, which will suit Jacobi's method. Then $\frac{z'}{z} = \frac{q}{p}$, and the term of the second order in δS becomes

$$\frac{\pi}{2}\int \frac{y}{(1+p^2)^{\frac{3}{2}}}\left(\delta p - \frac{q}{p}\,\delta y\right)^2 dx\,;$$

and this is essentially positive.

108. The conclusion of the whole investigation is that we are sure of a minimum in some cases, and no argument can be drawn from Jacobi's method to shew that we fail to secure a minimum in any case.

109. It is easy to verify that $z = p$ satisfies the differential equation of Art. 23.

Put p for z; thus we get $Pp - \frac{d}{dx}(Qq)$, that is

$$-\frac{2p}{a} - \frac{2pq}{(1+p^2)^{\frac{3}{2}}} - \frac{d}{dx}\,\frac{2yq}{(1+p^2)^{\frac{3}{2}}}\,:$$

we must shew that this expression vanishes.

Now $$\frac{2ay}{(1+p^2)^{\frac{1}{2}}} = y^2 + c_1\,;$$

therefore $$\frac{2ap}{(1+p^2)^{\frac{1}{2}}} - \frac{2apqy}{(1+p^2)^{\frac{3}{2}}} = 2yp\,;$$

therefore $$\frac{qy}{(1+p^2)^{\frac{3}{2}}} = -\frac{y}{a} + \frac{1}{(1+p^2)^{\frac{1}{2}}}\,:$$

hence by differentiating we have the required result.

We might attempt to get the complete value of z from the differential equation, as we thus know a particular value. The differential equation is, by Art. 23,

$$Pz - \frac{d}{dx}\left(Q\frac{dz}{dx}\right) = 0;$$

that is, $$Pz - \frac{dQ}{dx}\frac{dz}{dx} - Q\frac{d^2z}{dx^2} = 0.$$

Assume $z = pv$; then since we know that

$$Pp - \frac{dQ}{dx}\frac{dp}{dx} - Q\frac{d^2p}{dx^2} = 0,$$

we get $$\frac{dv}{dx}\left(p\frac{dQ}{dx} + 2Q\frac{dp}{dx}\right) + Qp\frac{d^2v}{dx^2} = 0.$$

Here Q stands for $\frac{2y}{(1+p^2)^{\frac{3}{2}}}$.

Thus $$\frac{\frac{dQ}{dx}}{Q} + \frac{2\frac{dp}{dx}}{p} + \frac{\frac{d^2v}{dx^2}}{\frac{dv}{dx}} = 0;$$

therefore $$\frac{dv}{dx}Qp^2 = \text{constant} = 2C \text{ say};$$

therefore $$v = 2C\int\frac{dx}{Qp^2} = C\int\frac{(1+p^2)^{\frac{3}{2}}}{p^2y}dx.$$

This will be seen to be just equivalent to what we found before in Art. 102.

[It may be expedient to draw attention to the precise enunciation of the problem which has been discussed in Arts. 79...94 and 101...108. We have to determine the solid of revolution of minimum surface and given volume, under certain conditions: see Arts. 80 and 81. The condition, that the generating curve is never to be convex to the axis of revolution, is practically equivalent to the condition, that the generating curve is always to be concave to the axis of revolution: it is this condition which constitutes the difficulty and the interest of the discussion.

At the period of the controversy to which I have referred in Art. 73, the Rev. Joseph Horner of Clare College, to whom I have been frequently indebted for valuable communications on other branches of mathematics, suggested to me to undertake the investigation of the problem with this condition of concavity. I convinced myself then that the solution must consist of some combination of straight lines with a curve; but I did not obtain the definite result until I returned to the subject for the purpose of the present essay.

Some further remarks on the problem, which complete the application of Jacobi's method to it, will be found in Art. 288.]

CHAPTER VI.

PROBLEMS ANALOGOUS TO THAT IN CHAPTER V.

110. I WILL now take another problem which presents points of interest similar to those of the preceding problem, but which involves less complicated analysis.

111. The volume of a solid of revolution which cuts the axis at two fixed points A and B is given: determine the form of the

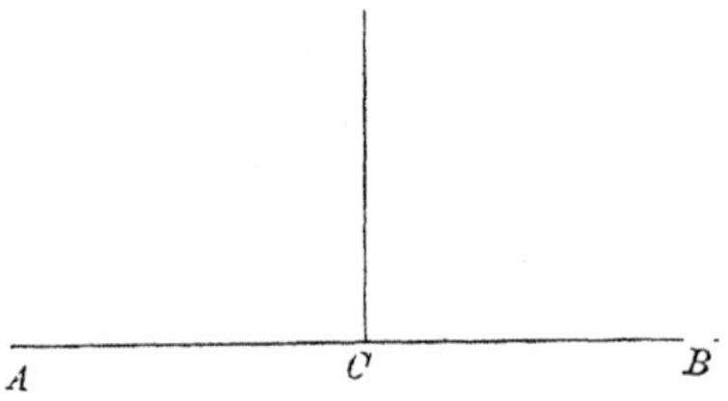

solid so that the moment of inertia round an axis at right angles to AB through C the middle point of AB may be a minimum.

Take C for the origin, and CB for the axis of x. Then the moment of inertia of the solid round the axis of y is

$$\pi \int \left(\frac{y^2}{4} + x^2 \right) y^2 \, dx,$$

and the volume is $\pi \int y^2 \, dx$; the limits of integration being the values of x at A and B respectively.

Hence by the usual theory we have to find the minimum of

$$\int \left(\frac{y^4}{4} + y^2 x^2 - a^2 y^2 \right) dx,$$

the limiting values of the variables being fixed, and a being a constant at present undetermined.

Denote the integral by u; then to the first order

$$\delta u = \int y\,(y^2 + 2x^2 - 2a^2)\,\delta y\,dx.$$

Thus we obtain as the necessary solution

$$y\,(y^2 + 2x^2 - 2a^2) = 0.$$

The interpretation is similar to that given in Art. 73. The solution consists of an oblate spheroid found by the equation

$$y^2 + 2x^2 - 2a^2 = 0,$$

which is connected by a straight line, namely, part of the axis of revolution, with the fixed points. The constant a must be determined so that the volume of the spheroid may have the given value. For facility of conception we may consider the connecting straight line as an infinitesimally slender cylinder.

There is however this difference: in the present case on account of the factor y the value of δu to the first order is always zero, for the discontinuous solution, instead of being a positive quantity of the first order: see Art. 74.

112. The term of the *second* order in the value of δu is

$$\int \left(\frac{3}{2}y^2 + x^2 - a^2\right)(\delta y)^2\,dx.$$

For the part of the solution which consists of the oblate spheroid this reduces to

$$\int y^2\,(\delta y)^2\,dx,$$

because here $\dfrac{y^2}{2} + x^2 - a^2 = 0$.

For the other part of the solution the term of the second order reduces to

$$\int (x^2 - a^2)\,(\delta y)^2\,dx,$$

because here $y = 0$;

and in this case x^2 is greater than a^2; hence the aggregate term of the second order in δu is necessarily positive; so that we have a minimum as required.

And as there must be a *least* value, and no other minimum value can be found than that here given, we may be sure that we thus obtain the least value.

113. If the given volume be $\frac{8\pi}{3} AC^3$ the solution consists of the oblate spheroid alone, and there is no discontinuity. But if the given volume have not this value there will be discontinuity.

If the given volume be less than $\frac{8\pi}{3} AC^3$ the solution resembles that of the lower diagram of Art. 76. If the given volume be greater than $\frac{8\pi}{3} AC^3$ the solution resembles that of the upper diagram of Art. 76; unless indeed there is the condition that the solid may not stretch beyond the straight lines drawn at right angles to the axis drawn through A and B respectively. This case we will now consider.

114. The following will be the solution in this case:

AD and BE are straight lines through the fixed points at right

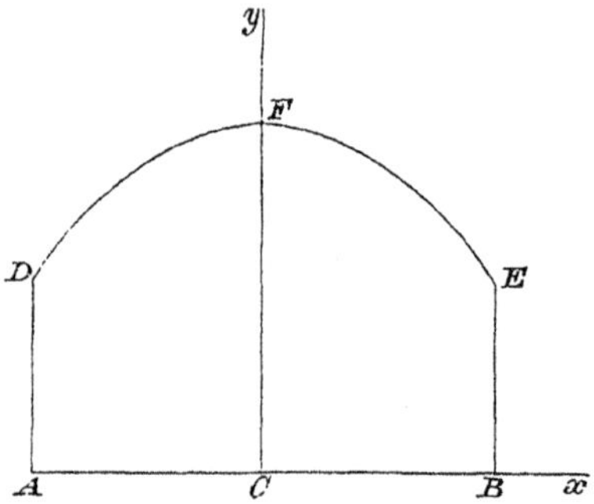

angles to the axis. DFE is an arc of an ellipse given by

$$y^2 + 2x^2 = 2a^2.$$

The required generating curve consists of AD, DFE, and EB. The constant a must be determined so that the volume generated by the revolution of this figure may have the assigned value.

To justify this solution all that is necessary is to observe that the term of the first order in the variation of the moment of inertia is zero, and the term of the second order is positive.

115. In the problem discussed in Arts. 73...78 we were able to confirm the proposed solution by appealing to the admitted fact that a sphere is the figure of greatest volume within a given surface. In the present problem we can confirm the proposed solution in an equally decisive manner.

Suppose an element of area indefinitely small in every direction situated at the point (x, y); let this generate a ring by revolving round the axis of x. Let m denote the mass of this ring; then the moment of inertia of the ring round the proposed axis is $m\left(\frac{y^2}{2}+x^2\right)$. Hence the solid of least moment of inertia with a given volume cannot be generated by any *curve* except the curve $\frac{y^2}{2}+x^2=\text{constant}$. For if there were any other generating curve we could obtain a less moment of inertia by removing matter which might be *outside* a bounding surface corresponding to this equation, and putting such matter *inside* the surface. The method here indicated has been already applied to problems relating to solids of greatest attraction: see Todhunter's *History of the Calculus of Variations*, Arts. 322 and 423.

116. It may be objected to the solutions of the present Chapter that our boundary has abrupt changes of direction, and a question may be raised as to the possibility of finding a solution which does not involve this abrupt change. I say that it is hopeless to seek for such a solution; the reasons for this assertion have been already stated: see Art. 14, Remark III.

117. Let us proceed now to consider the problem with conditions explicitly imposed like those in Art. 80. We suppose that the given volume is less than $\frac{8\pi}{3}AC^3$; and we require the solid of revolution to have this volume and a minimum moment of inertia, the generating line being never convex to the axis of revolution, and having no abrupt changes of direction. As in Art. 79 I venture to request any reader of these researches to investigate the problem for himself before examining the solution which I shall now propose.

118. As in Art. 82 every kind of boundary of the generating figure is excluded except straight lines and the ellipse

$$y^2 + 2x^2 = \text{constant}.$$

On trying various combinations I arrive at the conviction that there is no solution which exactly corresponds to the enunciation; but there is a limiting form to which we can approach as closely as we please: that is, we can find solids fulfilling all the required conditions, and having their moments of inertia never less than, though only infinitesimally greater than, a certain definite value.

119. Let AP and BQ be equal straight lines, making equal angles with AB. Let PDQ be an arc of an ellipse $y^2 + 2x^2 = 2a^2$.

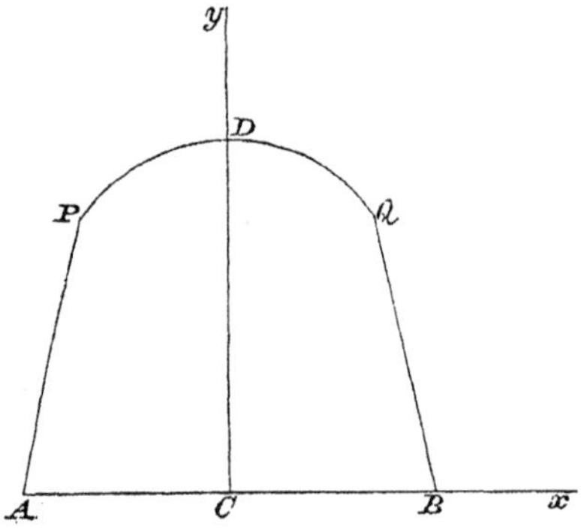

Let the constant a, and the angle PAC be determined so as to make the volume generated by the revolution of $APDQB$ round the axis of x equal to the assigned volume, and also to make the moment of inertia of the solid round the axis CD the least possible: this of course is only an ordinary problem in the Differential Calculus which we may assume to be capable of solution.

Then I say that the boundary thus determined is the limiting form of the solution to our problem in the Calculus of Variations. It does not strictly fulfil the conditions because there are abrupt changes of direction at P and Q; but we may suppose a curve drawn indefinitely near to this boundary so as to avoid the abrupt change of direction, and to be *always concave* to the axis. See the diagram to Art. 14.

120. We proceed to justify the statement of the preceding Article. We have to the first order

$$\delta u = \int y\,(y^2 + 2x^2 - 2a^2)\,\delta y\,dx,$$

and as $y^2 + 2x^2 - 2a^2 = 0$ for the arc PDQ this reduces to

$$\int y\,(y^2 + 2x^2 - 2a^2)\,\delta y\,dx,$$

where the integral is to be taken from the value of x corresponding to A to the value of x corresponding to P, and then from the value of x corresponding to Q to the value of x corresponding to B.

Now if δy be proportional to y this expression for δu will vanish; for by supposition AP and BQ are so taken as to make the moment of inertia a minimum corresponding to the assigned volume, and therefore of course an infinitesimal change in the position of AP and BQ must leave the value of u unchanged to the first order. In fact instead of saying that a and the angle PAC are to be determined so as to make the moment of inertia a minimum for the assigned volume, we might say that they are to be determined so as to produce the assigned volume and to make

$$\int y\,(y^2 + 2x^2 - 2a^2)\,\delta y\,dx$$

vanish between limits corresponding to AP and QB when δy is proportional to y; that is,

$$\int y^2\,(y^2 + 2x^2 - 2a^2)\,dx$$

must vanish between the limits corresponding to AP and QB.

121. We remark in passing that since the last integral is to vanish the expression $y^2 + 2x^2 - 2a^2$ must change its sign within the range of integration; this shews that if the arc PDQ were continued it would *cut* the straight lines AP and BQ. Hence it follows that the tangent at P to the arc PDQ is inclined to the axis of revolution at a less angle than AP is. This is essential to our solution, in order that when we draw a curve close to the boundary $APDQB$, as explained in Art. 119, this curve may always be *concave* to the axis.

122. The demonstration that we have a minimum is similar to that in Art. 85.

The straight line AP is supposed to be the same as in Art. 119; and ALK is a curve obtained from AP by ascribing admissible values to δy; so that the curve has no convexity towards the axis of revolution, and no abrupt change of direction.

The point F is not found by making $AF = \frac{2}{3} AP$ as in Art. 85; but F is the point where the straight line AP is cut by the curve $y^2 + 2x^2 = 2a^2$.

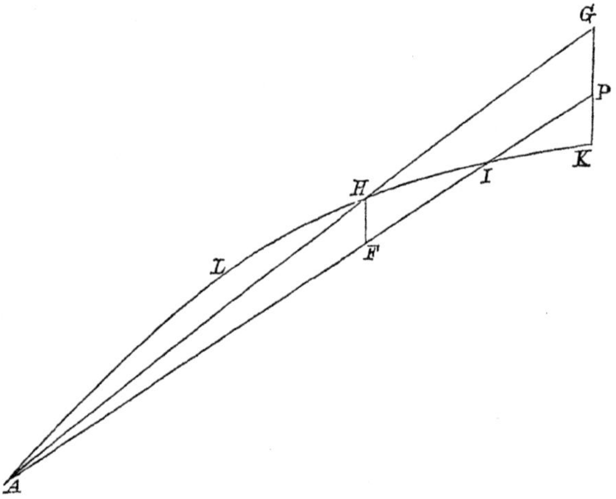

Thus in passing from AP to the curve $ALHK$ we have ultimately a gain of positive elements corresponding to the area ALH, a relief from the negative elements corresponding to the area $HIPG$, and a gain of positive elements corresponding to the area IKP. In like manner the other two diagrams of Art. 85 apply here.

123. Hence our conclusion is like that in Art. 86. We find that δu is always a positive quantity of the first order, and so we are sure of a minimum; except in one particular case, namely, that in which for the rectilinear part of the boundary δy is taken proportional to y, and then δu is zero to the first order.

In this particular case then we must examine the term of the second order in δu; as we have seen in Art. 112, this term is

$$\int\left(\frac{3y^2}{2}+x^2-a^2\right)(\delta y)^2\,dx.$$

Now for the part PDQ of the boundary this reduces to

$$\int y^2\,(\delta y)^2\,dx,$$

which is positive. For the parts AP and BQ we have by supposition δy proportional to y, so the sign of the term is the same as that of

$$\int\left(\frac{3y^2}{2}+x^2-a^2\right)y^2\,dx,$$

and therefore is the same as that of

$$\int y^2 \,.\, y^2\,dx;$$

for we know that for the limits with which we are concerned

$$\int\left(\frac{y^2}{2}+x^2-a^2\right)y^2\,dx=0.$$

Hence the term of the second order is positive.

Thus we can assert confidently that the proposed solution is a *minimum*.

And as there must be a *least* value, and no other minimum value presents itself besides this, we may conclude that this is also the *least* value.

124. In the discussion of the present Chapter we have the discontinuity of a straight line and a curve which meet without touching. As in Art. 100 we see that the discontinuity arises from a condition implicitly imposed, or explicitly imposed.

125. We may give a still more simple example of the kind discussed in the present Chapter. Find a curve which shall cut the axis of x at given points, and enclose an assigned area, and have a minimum moment of inertia round an axis at right angles to the plane of the curve passing through the point on the axis of x which is midway between the two given points.

Take the axes as in Art. 111; then proceeding in the usual way we find that we require the minimum of

$$\int\left(\frac{y^2}{3}+x^2-a^2\right)y\,dx.$$

Denote this by u; then to the first order

$$\delta u=\int(y^2+x^2-a^2)\,\delta y\,dx.$$

Thus we obtain as the necessary solution

$$y^2+x^2-a^2=0.$$

This is the equation to a circle having its centre at the origin. We obtain the same kind of discontinuity as we had in Arts. 73 and 111. The present case resembles that of Art. 73 in the circumstance that δu instead of being zero to the first order for the discontinuous solution is a positive quantity.

The investigations of Arts. 113...123 apply with obvious modifications to the present problem. The simplicity of the present problem recommends it as offering an easy case for the consideration of any person who might wish to contest the conclusions at which we have arrived.

126. Some variety might be introduced into the problems of Arts. 111 and 125 by giving a different position to the axis about which the moment of inertia is required. Instead of cutting the plane of x, y at the origin let this axis cut the plane at the point (h, k). Then in Art. 111 we have

$$u=\int\left\{\frac{y^2}{4}+(x-h)^2+k^2-a^2\right\}y^2\,dx;$$

and in Art. 125 we have

$$u=\int\left\{\frac{y^2}{3}+(x-h)^2+k^2-a^2\right\}y\,dx.$$

For in the latter case we may consider separately the parts of the figure which are above and below the axis of x; as it is obvious that each part must have the minimum property.

The centre of the ellipse in Art. 111, and the centre of the circle in Art. 125 is now not at the origin but at the point $(h, 0)$. It is easy to see what effect this will produce on our solution.

127. Suppose another condition imposed besides those in Art. 117: let the generating curve be required to have no abrupt change of direction, to be never convex to the axes, and to cut the axis at given points and *at given angles.*

For simplicity suppose these angles equal and denote them by γ.

Let the problem *as enunciated in Art.* 117 be solved; and suppose that the boundary there obtained cuts the axis at an angle β.

I. If $\beta = \gamma$ the solution of the problem in Art. 117 obviously satisfies all the conditions of the problem of the present Article.

II. If β is less than γ this solution must still be made to suffice. In the diagram of Art. 119 we must conceive straight lines making an angle γ with AB to be drawn, one through a point to the right of A but indefinitely close to A, and the other through a point to the left of B but indefinitely close to B.

Thus we obtain a boundary illustrated by the diagram; AE and BF will be indefinitely short. As in Arts. 118 and 119 the

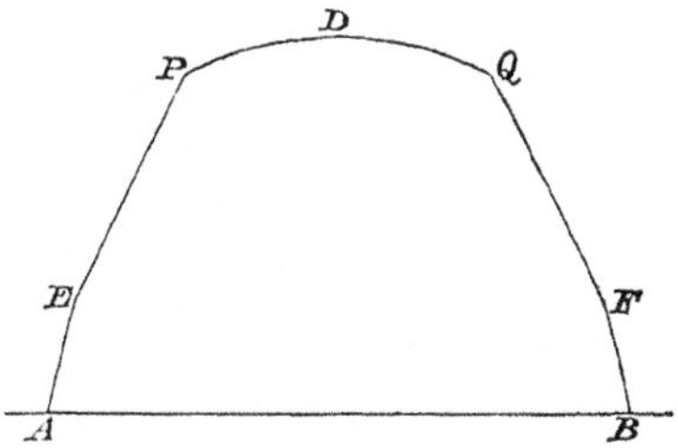

problem most strictly speaking does not admit of a solution. But the diagram gives the *limit* towards which we must approach as we make the moment of inertia continually smaller while we retain the conditions imposed at the commencement of this Article.

III. If β is greater than γ the solution of the problem in Art. 117 will be of no use in the present case. Our solution will then be as follows: Through A and B draw straight lines inclined at an angle γ to AB. Let AP and BQ denote these straight lines. Let PDQ be an arc of the ellipse $y^2 + 2x^2 = 2a^2$; and let a be so taken that the volume generated by the revolution of $APDQB$ round AB may have the assigned value.

To shew that this is a solution all that is necessary is to examine the value of δu.

Let $P'D'Q'$ be a curve obtained by variation from PDQ, and

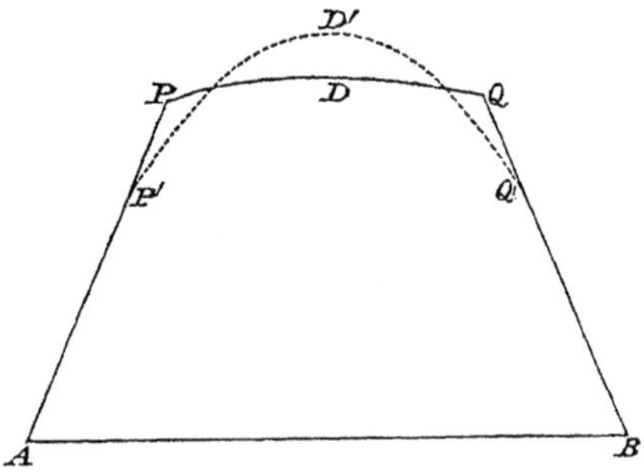

suppose that P' and Q' are below P and Q respectively. To the first order

$$\delta u = \int y\,(y^2 + 2x^2 - 2a^2)\,\delta y\,dx.$$

Now for limits corresponding to P and Q this vanishes. For the small portion between P and P', and that between Q and Q', we have δy negative and $y^2 + 2x^2 - 2a^2$ negative; so that there is a small *positive* element of δu: this would vanish if AP touched PQ at P. The term of the second order in δu is positive as in Art. 111.

If the curve obtained by variation has its ends *above* P and Q respectively, then δu to the first order vanishes entirely, and the term of the second order is positive as before.

Thus we have a minimum. Of course the given volume must be less than that of the double cone, which would be obtained by revolving round AB the triangle formed by producing AP and BQ to meet.

128. The discussion in the preceding Article suggests to us to try the effect of imposing another condition on the problem of the preceding Chapter; thus we have the following enunciation: determine the solid of revolution of minimum surface, and of given volume, supposing that the generating line is to cut the axis at given points, at given angles, and is never to be convex to the axis, and to have no abrupt change of direction.

129. For simplicity suppose the given angles to be equal, and denote them by γ.

Let the problem as enunciated in Art. 83 be solved; and suppose that the boundary thus obtained cuts the axis at an angle β.

I. If $\beta = \gamma$ the solution of the problem enunciated in Art. 83 obviously satisfies all the conditions of the problem of the present Article.

II. If β is less than γ the solution must still be made to suffice: the mode in which this must be effected is the same as that in Case II. of Art. 127.

III. If β is greater than γ the solution of the problem in Art. 83 will be of no use. I propose the following as the solution: Take a curve determined by $\dfrac{2ay}{\sqrt{(1+p^2)}} = y^2 + c$, where a and c are constants; and draw tangents at the points where the tangents will be inclined at an angle γ to the axis of x: then there are two conditions which must serve to find the values of the constants; namely, the tangents must intersect the axis of x at the given points, and the solid formed by the revolution of the boundary round the axis of x must have the assigned value.

It remains then to shew that these conditions can be satisfied, as well as the condition necessary for a minimum.

130. Let S denote the surface of the solid generated in the manner described; then we must shew that δS to the first order is zero or positive: this can be done in the manner already exemplified.

As in Art. 84 we can shew that

$$\delta S = 2\pi \int \left\{ \frac{1}{\sqrt{(1+p^2)}} - \frac{y}{a} \right\} \delta y \, dx,$$

where the integral is to be taken between limits corresponding to the rectilinear parts of the boundary: for these parts we have

$$\frac{1}{\sqrt{(1+p^2)}} = \cos\gamma.$$

Now the condition that there is to be no convexity in fact renders δy necessarily negative as far as we are concerned with it.

If the ordinate at the points common to the rectilinear and curvilinear parts is $\frac{3a}{2}\cos\gamma$, we may by taking δy proportional to y as an extreme admissible case just make δS zero to the first order.

If the ordinate at the specified points is greater than $\frac{3a}{2}\cos\gamma$, then a gain of positive elements is secured, and δS is positive to the first order.

Then if δy be taken in any other admissible way, δS is positive to the first order: see the reasoning and the diagram of the last of the three cases of Art. 86.

Thus δS is zero or positive to the first order provided the ordinate at the point of discontinuity is not less than $\frac{3a}{2}\cos\gamma$.

If the ordinate is *greater* than $\frac{3a}{2}\cos\gamma$, we are sure of a minimum even without looking at the terms of the second order.

131. Now let us consider the conditions which the constants must satisfy. Put the equation to the curve in the form

$$\frac{2ay}{\sqrt{(1+p^2)}} = y^2 + a^2(1-e^2),$$

so that a and e are the constants.

Let y_1 be the ordinate of the point which is common to the rectilinear and the curvilinear part of the boundary. Then, by Art. 90, we have

$$y_1 = a\cos\gamma + a\sqrt{(e^2-\sin^2\gamma)} \ldots\ldots\ldots\ldots (1);$$

and this is not to be less than $\frac{3a}{2}\cos\gamma$;

therefore $\sqrt{(e^2-\sin^2\gamma)}$ not less than $\frac{1}{2}\cos\gamma$;

therefore e^2 not less than $\sin^2\gamma + \frac{1}{4}\cos^2\gamma$.

Let $2h$ denote the distance between the given points; then

$$h = y_1\cot\gamma + a\int_0^{\tan\gamma} L\,dp \ldots\ldots\ldots\ldots (2),$$

where L has the meaning assigned in Art. 90.

Let $2V$ denote the volume generated by the revolution of the boundary round the axis of x; then

$$V = \frac{\pi y_1^3\cot\gamma}{3} + \pi a^3\int_0^{\tan\gamma} H\,dp \ldots\ldots\ldots\ldots (3),$$

where H has the meaning assigned in Art. 91.

Suppose the value of y_1 from (1) substituted in (2) and (3); then from (2) we find $\frac{a}{h}$ = a function of e and known quantities. Substitute this value of a in (3); thus V becomes a function of e and of known quantities. And as e changes continually from an indefinitely large value we see that V will vary continually, as long as e is greater than $\sin\gamma$.

At the extreme case of e infinite we have a indefinitely small; and $ae \cot \gamma = h$. Then $V = \frac{\pi y_1^3 \cot \gamma}{3}$ where $y_1 \cot \gamma = h$.

The assigned volume must, of course not exceed $\frac{2\pi h^3 \tan^2 \gamma}{3}$ or the problem will be impossible.

At the value $e = \sqrt{\sin^2 \gamma + \frac{1}{4} \cos^2 \gamma}$ the value of V will be the same as we should obtain in Art. 91 if β were changed into γ. Now this value is less than the assigned value of the present problem; for this assigned value corresponds to that obtained in Art. 91 with the value β which is greater than γ: and we know that the value of V in Art. 91 increases with e, and is therefore greater for β than for γ.

Hence finally as from (3) the value of V ranges between limits which include between them the assigned value of the present problem, it will be possible to take e such that V shall just be equal to the assigned value.

132. We may feel sufficiently secure that there is only one solution thus: The greatest possible assigned volume corresponds to e infinite and a indefinitely small; then if $\frac{dV}{de}$ be the differential coefficient of V with respect to e when a is supposed eliminated we are sure that $\frac{dV}{de}$ is positive when e is infinite. Therefore if e be diminished V diminishes. Suppose that $\frac{dV}{de}$ could vanish at any point; then we should have the same value of V corresponding to two adjacent values of e. Thus we get two adjacent solutions. But this is impossible; for by Art. 130 each is a true minimum, therefore each is less than the other.

133. We may remark a difference between this problem and that in Art. 117. Here the rectilinear and curvilinear parts *touch*, while there they met at a finite angle. The difference depends

on the circumstance that in the present problem the expression under the integral sign in the quantity to be a minimum involves the differential coefficient p, which did not occur in the former problem.

134. Another modification may be given to the problem of the preceding Chapter. A bowl is to be made in the form of a figure of revolution so as to have a given surface and to hold a maximum volume.

If the problem be stated thus without any condition the bowl will have to be a sphere; this must be considered a limiting case, as strictly speaking a closed surface cannot be called a bowl.

But suppose the breadth of the bowl at the open end to be given.

Let Ox be the axis of revolution; let $2k$ denote the given breadth, and suppose $OB = k$.

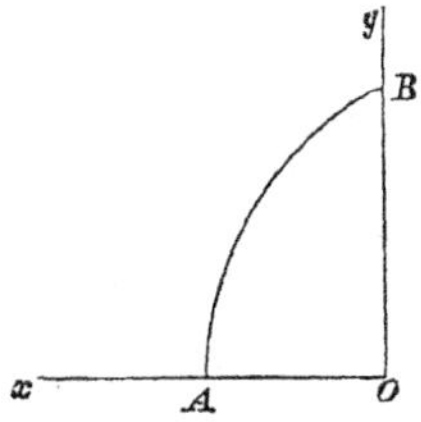

The solution, as in Art. 73, is a circle which has its centre on the axis of x. Let the radius of the circle be r, and let OA, the depth of the bowl, $= h$. Then to find h and r we have the equations

$$k^2 = (2r - h)\, h,$$

$$2\pi rh = \text{the given surface} = S \text{ say};$$

therefore $$h^2 = \frac{S}{\pi} - k^2.$$

This gives always a real value for h, as S must of course be not less than πk^2.

Next suppose that the depth of the bowl is also given.

If this given depth is greater than that which our solution assigns, we must take this solution and suppose an indefinitely

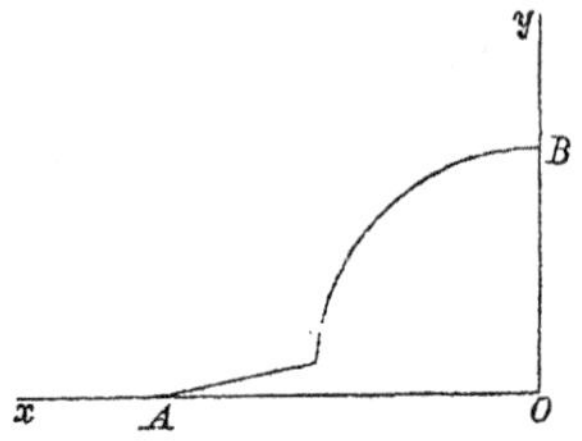

narrow conical or cylindrical part attached at the bottom of the bowl, and having the same axis as the bowl.

If the given depth is less than that which our solution assigns we may understand the condition in two ways. If we merely mean that the depth as measured *along the axis* is to have the given value, we use such a solution as arises from having the point of the conical part in the diagram inwards instead of outwards. But if we mean that the depth is at no point to be greater than the given quantity, then the solution must be composed partly of a straight line through A at right angles to the axis of x, and partly of a curve passing through B and touching this straight line, and satisfying the differential equation

$$\frac{2ay}{\sqrt{(1+p^2)}} = y^2 + b\,;$$

see Art. 73.

135. A curve generates a bowl by revolving through 180° round an axis which it cuts at two fixed points: find the curve so that the centre of gravity of the surface may be at the greatest distance from the axis, the area of the surface being given.

The enunciation does not immediately suggest any difficulty or strangeness as likely to occur in the solution.

We have by the usual theory to find a maximum of

$$\int (y^2 - ay) \sqrt{(1+p^2)}\, dx.$$

Call it u; then we obtain

$$\frac{y^2 - ay}{\sqrt{(1+p^2)}} = \text{constant};$$

the constant must be zero since the curve is to cross the axis.

We cannot take $y = a$, for then we shall not have $\delta u = 0$ to the first order, since

$$\delta u = \int \left\{ (2y-a)\sqrt{(1+p^2)} - \frac{d}{dx} \frac{p(y^2 - ay)}{\sqrt{(1+p^2)}} \right\} \delta y\, dx.$$

We must then attempt to form a solution by combining $p = \infty$ and $y = 0$; the former gives a straight line parallel to the axis of y, and the latter is the axis of x.

This indicates that strictly speaking there is no solution; but there is a kind of limit towards which we may approach

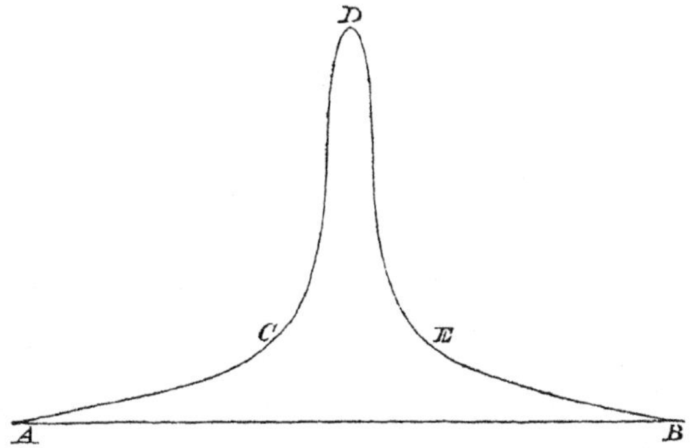

indefinitely. The generating curve must be supposed to run indefinitely close to the axis, except where it turns off, nearly at right angles to the axis, and returns again; thus $ACDEB$ may represent it, CD and DE being very close to each other, and nearly straight lines.

The limit to which we approach consists of the straight line AB and a straight line FD at right angles to AB; the position

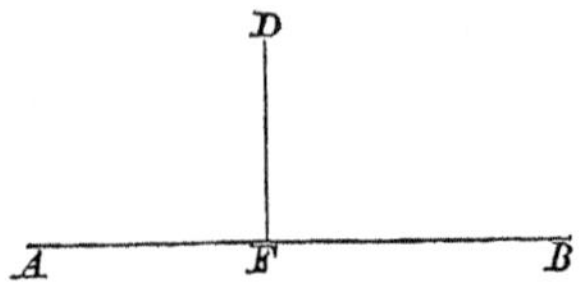

of F is arbitrary; the length of FD must be such that πFD^2 may be equal to half the given surface.

To shew that the limit is a true maximum, we may proceed as in Art. 64.

We may observe that instead of a single straight line FD at right angles to AB we might take two or more. Suppose we take two, say FD and HK. Then we must have

$$\pi (FD^2 + HK^2)$$

equal to half the given surface. This would still be a maximum. Instead of *one* long slender part CDE, as in the former figure, we should now have *two*. It is obvious however that the distance of the centre of gravity from the axis is greater when there is only one long slender part than when there are two or more such parts. For in passing from the former case to the latter, we in fact remove matter from one position and put it into another which is nearer to the axis; and thus we bring the centre of gravity nearer to the axis. We may of course obtain the same result by using the common formula for the position of the centre of gravity.

136. Suppose we ask for the solution of this problem with the condition that the boundary is never to be convex to the axis and to have no abrupt changes of direction.

Now, as in Art. 82, every kind of boundary of the generating figure is excluded by the Calculus of Variations, except straight lines, and curves which satisfy the differential equation

$$\frac{y^2 - ay}{\sqrt{(1+p^2)}} = \text{constant}.$$

Proceeding as in Art. 83, I suggest the following for the solution of the problem.

Let A and B be the fixed points on the axis; let AP and BQ

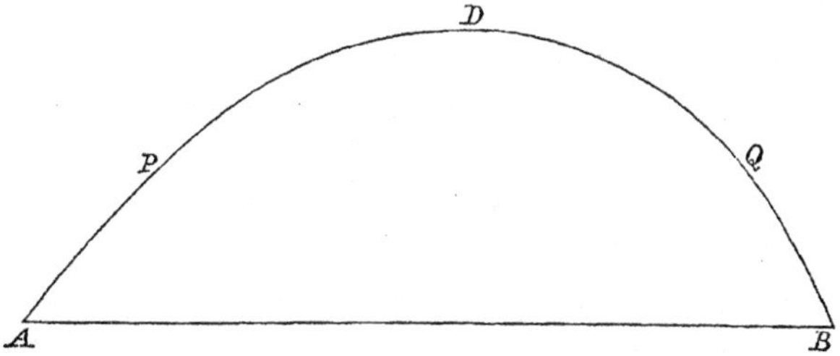

be equal straight lines equally inclined to the axis which touch the curve PDQ at P and Q respectively; and let PDQ be an arc of the curve determined by the above differential equation. The constants must be so taken as to ensure the tangency at P and Q, and to make the surface generated by $APDQB$ have the given value. Moreover there is a condition to be satisfied which we shall investigate presently; this condition connects the constant a with the ordinate of P and Q.

After the full investigation in Chapter V. it will not be necessary to discuss the present problem in detail. We see that the expression for δu given in Art. 135 vanishes for the part PDQ, and for the parts AP and QB it reduces to

$$\int \frac{2y - a}{\sqrt{(1+p^2)}} \delta y \, dx.$$

Let y_1 denote the ordinate of P and Q; then as in Art. 85 we see that y_1 must be such that

$$\frac{2y_1^3}{3} - \frac{ay_1^2}{2} = 0;$$

thus
$$y_1 = \frac{3a}{4}.$$

Let β denote the angle PAB or QBA; then if we put $-c^2$ for the constant in the above differential equation we see that

$$\cos\beta = \frac{16c^2}{3a^2}.$$

At the highest point D we shall have

$$y^2 - ay + c^2 = 0;$$

so that
$$y = \frac{a \pm \sqrt{(a^2 - 4c^2)}}{2}:$$

it will be found on investigation that we must take the upper sign.

137. Given the mass of a solid of revolution of uniform density, required its form so that the attraction on a point in the axis may be a maximum.

This is a well-known problem. Taking the origin at the point, and the axis of revolution for the axis of x we require the maximum of

$$\int\left\{1 - \frac{x}{\sqrt{(x^2+y^2)}} + ay^2\right\} dx,$$

where a is a constant.

Denote the integral by u; then to the second order

$$\delta u = \int\left\{\frac{xy}{(x^2+y^2)^{\frac{3}{2}}} + 2ay\right\}\delta y\, dx + \int\left\{a + \frac{x(x^2-2y^2)}{2(x^2+y^2)^{\frac{5}{2}}}\right\}(\delta y)^2\, dx.$$

Thus the solution is to be found from

$$\frac{x}{(x^2+y^2)^{\frac{3}{2}}}+2a=0\,;$$

it will be convenient to write $-\frac{1}{c^2}$ for $2a$; thus $(x^2+y^2)^{\frac{3}{2}}=c^2x$.

When $y=0$ we have $x=0$ or c; thus c is the length of the axis.

$$\text{The volume } =\pi\int_0^c y^2dx=\pi\int_0^c\{(c^2x)^{\frac{2}{3}}-x^2\}\,dx$$

$$=\pi c^3\left(\frac{3}{5}-\frac{1}{3}\right)=\frac{4\pi c^3}{15},$$

so that the volume being given c is known.

In the term of the second order in δu the coefficient of $(\delta y)^2$ becomes

$$-\frac{1}{2c^2}+\frac{x\,(3x^2-2c^{\frac{4}{3}}x^{\frac{2}{3}})}{2\,(c^2x)^{\frac{5}{3}}},$$

that is, $\quad -\dfrac{1}{2c^2}+\dfrac{(3x^{\frac{4}{3}}-2c^{\frac{4}{3}})}{2c^{\frac{10}{3}}}$, that is, $\dfrac{3\,(x^{\frac{4}{3}}-c^{\frac{4}{3}})}{2c^{\frac{10}{3}}}$;

so that the term is $\quad \dfrac{3}{2c^{\frac{10}{3}}}\displaystyle\int_0^c(x^{\frac{4}{3}}-c^{\frac{4}{3}})\,(\delta y)^2\,dx.$

This is negative as it should be.

It is well known from other considerations that we have really a maximum. [See Todhunter's *History of the Calculus of Variations*, Art. 322.]

138. Now let us impose the condition that the length of the axis shall have a given value. The solution already obtained will not apply; because in this solution the length of the axis is determined as soon as the volume is given.

We observe however that the term of the first order in δu vanishes if $y=0$; and this suggests a solution of the kind already exemplified; namely, the generating curve must be made up of $(x^2+y^2)^{\frac{3}{2}}=c^2x$, together with a portion of the axis of x.

Let h denote the given length of the axis; then the term of the second order in δu becomes

$$\int_0^c \frac{3\,(x^{\frac{4}{3}} - c^{\frac{4}{3}})}{2c^{\frac{10}{3}}} (\delta y)^2\, dx + \int_c^h \left(-\frac{1}{2c^2} + \frac{1}{2x^2}\right) (\delta y)^2\, dx.$$

The first term is negative, and so is the second whether c be less or greater than h. Thus we have a solution whether c be less or greater than h; the case in which c is greater than h resembling that which has already presented itself in Art. 134 with the point of the cone turned *inwards*. But if c be greater than h the problem may be understood in another sense, and perhaps more naturally: namely, that the solid is *not to extend beyond* the ordinate at the point $x = h$. The solution then is given by

$$(x^2 + y^2)^{\frac{3}{2}} = \gamma^2 x,$$

where the constant γ is to be determined from the condition $\pi \int_0^h y^2 dx =$ the given volume. All the circumstances of the problem are thus satisfied: it will be observed that there is no term outside the integral sign in the value of δu. The generating curve may be said to consist of the curve $(x^2 + y^2)^{\frac{3}{2}} = \gamma^2 x$ from $x = 0$ to $x = h$, and of the ordinate at the point $x = h$.

If any objection is taken to the abrupt change of direction at

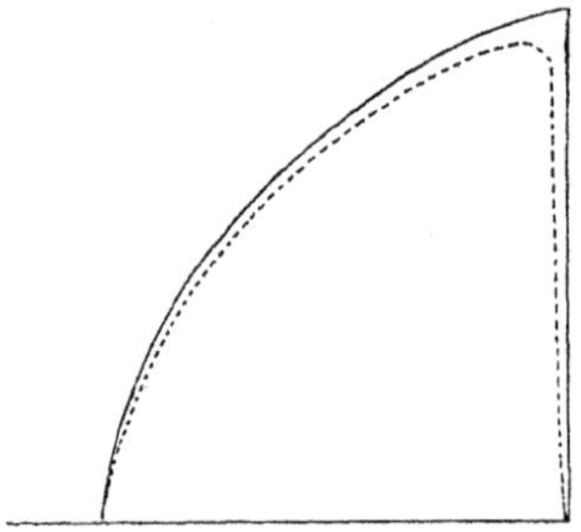

the point where the curve joins the straight line we must answer the objection in the same manner as before: see Art. 14.

139. Suppose we attempt to introduce also the condition that the radius of the base shall have a given value k, the length of the axis being h as before.

We shall have in some cases to employ the solution already given as a limiting form in the manner formerly exemplified.

Thus suppose $CA = h$, $CB = k$; and let the given volume be such that the curve ADE would have been the solution if h and k had not been mentioned. Then we must conceive a curve drawn indefinitely close to $ADECB$, and the closer it is drawn the nearer it will approach to the form which gives the greatest attraction under the specified conditions.

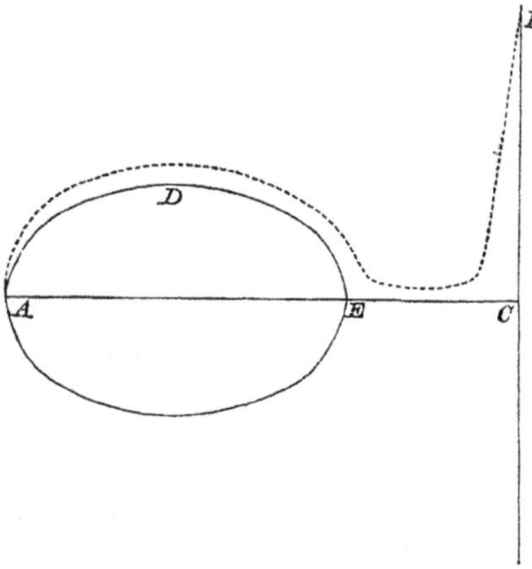

Or instead of the solution of Art. 137 we may have to take the solution of Art. 138 and make it in a similar manner applicable to the problem with the additional condition we have now imposed. There will be two varieties, because the base obtained in Art. 138 may be greater or may be less than that which is now prescribed, and thus there will be a difference in the mode of adjustment to the prescribed base.

It may happen that in these solutions some of the ordinates are greater than k.

But if we understand our condition to mean that we are to have no ordinate greater than k we shall require in some cases another solution, namely, a portion of the curve $(x^2+y^2)^{\frac{3}{2}} = \gamma^2 x$ combined with a portion of the straight line $y = k$.

The condition for finding γ will now be

$$\pi \int_0^{\xi} y^2 dx + \pi k^2 (h - \xi) = \text{the given volume,}$$

where ξ stands for AE, so that $(\xi^2 + k^2)^{\frac{3}{2}} = \gamma^2 \xi$.

Here δu to the first order vanishes with respect to AD; and

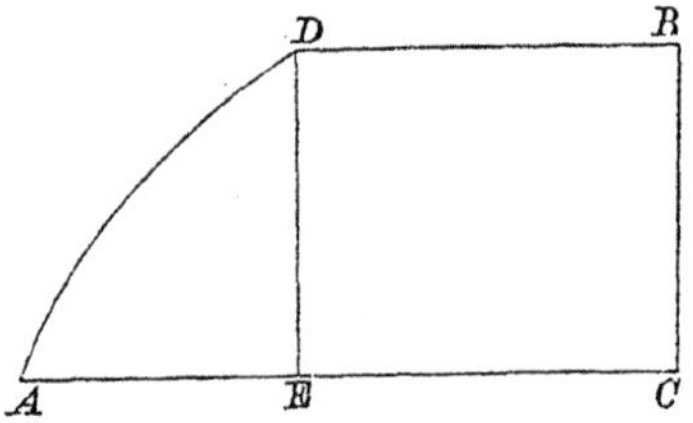

with respect to DB reduces to $\int_{\xi}^{h} \left\{ -\frac{k}{\gamma^2} + \frac{kx}{(x^2+k^2)^{\frac{3}{2}}} \right\} \delta y \, dx$,

that is to $$\frac{k}{\gamma^2} \int_{\xi}^{h} \left\{ \frac{\gamma^2 x}{(x^2+k^2)^{\frac{3}{2}}} - 1 \right\} \delta y \, dx.$$

Now if x lies between ξ and h the expression $\frac{\gamma^2 x}{(x^2+k^2)^{\frac{3}{2}}} - 1$ is positive; for this expression would be zero if we put instead of k the ordinate y of the curve corresponding to the abscissa x, and y is greater than k, so that the expression as it stands must be positive. And along DB we have δy essentially negative if it is not zero. Hence the above variation is a negative quantity of the first order unless δy vanishes at every point of DB. In the latter case we proceed to the term of the second order in the variation which reduces to $\int_0^{\xi} \frac{3(x^{\frac{4}{3}} - c^{\frac{4}{3}})}{2c^{\frac{10}{3}}} (\delta y)^2 \, dx$, which is negative. Thus a maximum is secured.

[We here confine ourselves to the case in which the curve AD, when continued beyond D, does not cut the straight line $y=k$ again between D and B. It is easy to see what modifications have to be made if the curve does cut this straight line again between D and B.]

140. The examples discussed in the present Chapter furnish numerous illustrations of the principle that discontinuity arises from conditions imposed in the problems. In Art. 139 we have a very simple example of the general principle of Art. 17.

CHAPTER VII.

BRACHISTOCHRONE UNDER THE ACTION OF GRAVITY.

141. In the present Chapter we shall discuss some cases of discontinuity which arise by imposing various conditions on the problem of determining the shortest course of a particle under the action of gravity.

142. Required the curve of quickest descent from a fixed point A to a fixed point B, supposing the moving particle constrained to remain on a fixed smooth inclined plane which contains A and B.

The curve is known to be a cycloid with its base horizontal, having a cusp at A; in fact we resolve the force of gravity into two components, one at right angles to the inclined plane, and the other along this plane: then the former component may be disregarded.

Now suppose we require the curve of quickest descent from A to B, when there are two fixed smooth inclined planes, one passing through A and the other through B, and the moving particle is constrained to move first on one plane, and then on the other, assuming that no velocity is lost in passing from one to the other.

The required curve consists of two arcs of cycloids with their bases horizontal; the first arc has a cusp at A; the second arc must have its cusp so situated in the second inclined plane that the velocity of the particle at the commencement of this arc is that which would be acquired in falling from the cusp: hence the cusp will in fact be in the horizontal plane which passes through A.

It is easily seen by examining the term in the variation which is outside the integral sign, that at the point where the two arcs of cycloids meet their tangents must be equally inclined to the line of intersection of the two planes.

143. A particle falls from a fixed point A to a fixed point B, passing through another point C: find the entire path when the time of motion is a minimum (1) supposing C to be a fixed point, (2) supposing C constrained to lie on a given curve.

[This problem was proposed by the present writer in the Mathematical Tripos Examination of 1866.]

We assume that no change of velocity occurs at the point C.

(1) From A to C the path must be a cycloid having its base horizontal and a cusp at A; from C to B the path must be a cycloid having its base horizontal and a cusp in the horizontal plane through A. This statement is evident from the combination of two known results; one relating to the brachistochrone when the initial velocity is zero, and the other relating to the brachistochrone when the initial velocity has a given value which is not zero. If C is not in the vertical plane which contains A and B, the two portions of the entire path are in different vertical planes.

(2) From A to C the path must be a cycloid having a cusp at A; from C to B the path must be a cycloid having a cusp in the horizontal plane through A: each cycloid has its base horizontal. This statement is evident for the same reason as before. Also the tangents to the two portions of the path which meet at C must make equal angles with the tangent to the given curve at C: this condition serves to determine the point C. To demonstrate this we proceed thus: Let us first suppose that A and B and the given curve are all in the same vertical plane. Let the axis of y be vertically downwards, and the axis of x horizontal. Let x and y denote the co-ordinates of C. Let $\psi'(x)$ denote the value of $\frac{dy}{dx}$ for the given curve at the point C; let p_0 and p_1 denote similar things for the two portions of the path which meet at C. Then, as in Todhunter's *Integral Calculus*, Art. 361, the part of the

variation outside the integral sign reduces to

$$\frac{1+p_1\psi'(x)}{\sqrt{y(1+p_1^2)}}\,dx - \frac{1+p_0\psi'(x)}{\sqrt{y(1+p_0^2)}}\,dx;$$

and in order that this may vanish we must have

$$\frac{1+p_1\psi'(x)}{\sqrt{(1+p_1^2)}} = \frac{1+p_0\psi'(x)}{\sqrt{(1+p_0^2)}}:$$

this involves the condition stated above.

If A and B and the given curve are not in the same vertical plane, the same condition may still be shewn to hold: see Art. 367 of Todhunter's *Integral Calculus*.

144. Required the curve of quickest descent from a fixed point A to a fixed point B; supposing that a screen or obstacle is interposed between A and B, having a given finite aperture through which the path must pass.

It may happen that the aperture is so situated that the cycloid of quickest descent from A to B passes through it: and then of course the condition imposed does not affect the solution of the problem. Suppose however that the aperture is not so situated, then, I say, that the required path must pass through some point of the *boundary* of the aperture. For if the path did not pass through the boundary there would be no limitation imposed on the dependent variable, and thus by the ordinary theory we are sure that we have not a curve of minimum time. Thus the path must pass through the *boundary* of the orifice; and therefore the present problem is reduced to that of Art. 143.

145. Again: find the curve of quickest descent from a fixed point A to a fixed point B with the condition that the path must pass through some point C which is on a given fixed surface.

It may happen that the given fixed surface is so situated that the cycloid of quickest descent from A to B crosses it; and then of course the condition imposed does not affect the solution of the problem. Suppose however that the given fixed surface is not so situated. The position of the point C must by Art. 143 be such that the two parts of the path make equal angles with the tangent to any curve drawn on the surface through the point of contact: the two parts must therefore make equal angles with the normal to the surface at that point, and must lie in the same plane with that normal.

To shew this however from the ordinary formulæ we proceed thus: Let p stand for $\frac{dy}{dx}$, and ϖ for $\frac{dz}{dx}$; these being formed on the curve. Let the subscript 1 apply to the end of the first portion of the path, and the subscript 2 to the beginning of the second portion. Let $\phi(x, y, z) = 0$ be the equation to the surface; put v^2 for $1 + p^2 + \varpi^2$. Then in order that the integrated part of the variation may vanish, we must have

$$\left\{\frac{p\delta y + \varpi\delta z}{v} + v dx\right\}_1 - \left\{\frac{p\delta y + \varpi\delta z}{v} + v dx\right\}_2 = 0 \;\ldots\ldots (1).$$

Now we must of course have the same point by variation, whether we consider (x, y, z) to be the end of the first portion of the path or the beginning of the second portion. Thus

$$\left.\begin{aligned} \delta y_1 + p_1 dx = \delta y_2 + p_2 dx = \delta\sigma \text{ say} \\ \delta z_1 + \varpi_1 dx = \delta z_2 + \varpi_2 dx = \delta\tau \text{ say} \end{aligned}\right\} \;\ldots\ldots\ldots\ldots (2).$$

Moreover, since the point obtained by variation is also on the given surface,

$$\frac{d\phi}{dx} dx + \frac{d\phi}{dy} \delta\sigma + \frac{d\phi}{dz} \delta\tau = 0 \;\ldots\ldots\ldots\ldots\ldots\ldots (3).$$

From (2) we can write (1), thus:

$$\frac{dx + p_1\delta\sigma + \varpi_1\delta\tau}{v_1} = \frac{dx + p_2\delta\sigma + \varpi_2\delta\tau}{v_2} \;\ldots\ldots\ldots\ldots\ldots (4).$$

Now regard dx, $\delta\sigma$, $\delta\tau$ as proportional to the direction cosines of a straight line. Then (3) shews that the straight line must be at right angles to the normal to the surface; and (4) shews that it must make equal angles with the tangents to the two portions of the path at the common point. And as the straight line may be any straight line at right angles to the normal, it is obvious that the tangents to the two portions of the path must satisfy the conditions stated above.

If we wish, however, to proceed in a more formal manner, we should substitute for dx from (3) in (4), and then equate the coefficients of $\delta\sigma$ and $\delta\tau$ to zero. Thus we get

$$\frac{1}{v_1}\left(p_1\frac{d\phi}{dx}-\frac{d\phi}{dy}\right)=\frac{1}{v_2}\left(p_2\frac{d\phi}{dx}-\frac{d\phi}{dy}\right);$$

$$\frac{1}{v_1}\left(\varpi_1\frac{d\phi}{dx}-\frac{d\phi}{dz}\right)=\frac{1}{v_2}\left(\varpi_2\frac{d\phi}{dx}-\frac{d\phi}{dz}\right).$$

Eliminate v_1 and v_2 by cross multiplication; and it will be found that we arrive at the condition which makes the normal and the two tangents lie in one plane.

And we can shew that *numerically*

$$\frac{1}{v_1}\left(\frac{d\phi}{dx}+p\frac{d\phi}{dy}+\varpi\frac{d\phi}{dz}\right)_1=\frac{1}{v_2}\left(\frac{d\phi}{dx}+p\frac{d\phi}{dy}+\varpi\frac{d\phi}{dz}\right)_2.$$

For if we substitute for $\frac{d\phi}{dy}$ and $\frac{d\phi}{dz}$ in terms of $\frac{d\phi}{dx}$ from the two equations just given, we shall find that each member numerically

$$=\frac{v_1v_2-(1+p_1p_2+\varpi_1\varpi_2)}{v_2-v_1}\frac{d\phi}{dx}.$$

Of course this equal inclination of the tangents to the normal to the surface, and their lying in the same plane with it, might have been anticipated. For through an infinitesimal portion of the path we may consider that the velocity does not sensibly vary: and so, by a well-known geometrical fact, the path should in the neighbourhood of the surface resemble that of a ray of light reflected at the surface.

Suppose we require the curve of quickest descent from a fixed point A to a fixed point B, with the condition that the path must not cross an obstacle in the shape of a fixed closed surface.

In this case, supposing the obstacle so situated that the ordinary cycloid is inapplicable, the solution will consist of three parts. The first part and the third part will be arcs of cycloids; the second part will be the brachistochrone for a particle constrained to move on the given surface, and we need not discuss this, for it is a well-known problem. The parts of which the solution consists will have common tangents at the points where they meet: this is obvious from what has already been given.

[In the original essay there was some confusion between the two problems of the present Article, the second problem being enunciated and the first solved: hence some slight changes in the Article were necessary, and have been made.]

146. The problems which we have hitherto considered in this Chapter illustrate the principle that discontinuity arises from conditions imposed. The discontinuity here is that the first and the last parts of the path are arcs of different cycloids, generally in different planes, and meeting at a finite angle: in the second problem of Art. 145 there is however an intermediate arc between the two arcs of cycloids.

It is obvious that in all the problems of this Chapter there must exist some solution which gives the *least* time of motion; and the Calculus of Variations shews that no other solution can exist than that which we have taken in each case respectively. Hence we need not consider the terms of the second order in the variation; in fact, however, in all the cases except the first of Art. 143 we should find that the sign of the term of the second order cannot be ascertained by any theory at present known.

147. Required the curve of quickest descent from a fixed point A to a fixed point B; with the condition that the particle is never to descend below the horizontal straight line through B.

By the principles of the Calculus of Variations combined with our condition every line is excluded except a cycloid having its base horizontal and its cusp at A, and the horizontal straight line through B.

If then the horizontal distance between A and B does not exceed $\frac{\pi}{2}$ times the vertical distance, the required line consists of the cycloid alone: our condition does not come into operation.

If the horizontal distance between A and B exceeds $\frac{\pi}{2}$ times the vertical distance, the required curve consists of the cycloid AC

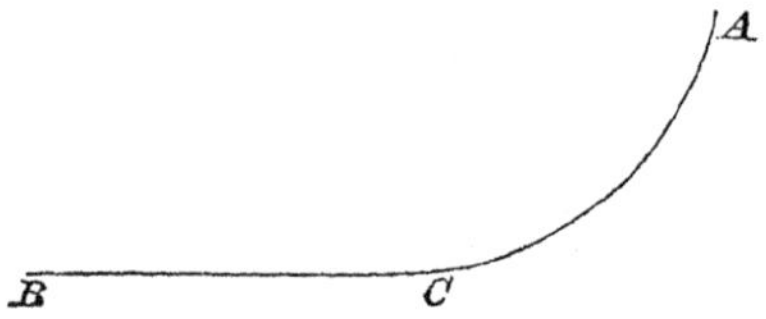

and the horizontal straight line CB, which is the tangent to the cycloid at its vertex C.

148. The example in the preceding Article resembles others which have already been discussed: the boundary which determines the area into which the path is not to pass may itself form through some of its extent a part of the required solution.

If the boundary instead of a *horizontal* straight line through B is any straight line, the solution is substantially the same.

If a straight line through A is the boundary which the path is not to pass, the required line consists of AC, a portion of this

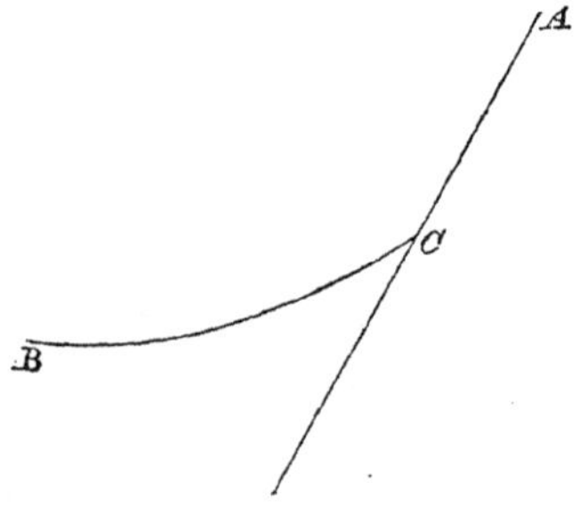

straight line, and CB an arc of a cycloid; the cycloid must have its base horizontal and its cusp in the horizontal straight line through A, and must touch the bounding straight line at C.

If the boundary is a straight line which does not pass through A or B, and the single cycloid from A to B is inapplicable because it would cross the boundary, the solution consists of three parts: an arc of a cycloid having its cusp at A and touching the boundary; a part of the boundary; and an arc of a cycloid which touches the boundary, passes through B, and has its cusp in the horizontal straight line through A: each cycloid has its base horizontal.

149. Required the curve of quickest descent from a fixed point A to a fixed point B, with the condition that the path is not to pass outside the given circular arc AB which does not exceed a quadrant, B being the lowest point of the circle.

It will be convenient to work out the solution from the beginning.

Take any point on the horizontal straight line through A for origin. Let the axis of x be horizontal, and the axis of y vertically downwards. We have to find a minimum value of

$$\int \frac{\sqrt{1+p^2}}{\sqrt{y}}\,dx.$$

Call this u: then

$$\delta u = P\delta y + \int \left(N - \frac{dP}{dx}\right)\delta y\,dx,$$

where $$N = -\frac{\sqrt{1+p^2}}{2y^{\frac{3}{2}}}, \quad P = \frac{p}{\sqrt{y(1+p^2)}}.$$

So long as δy is susceptible of either sign we know that there cannot be a minimum unless $N - \dfrac{dP}{dx} = 0$; and this leads to a cycloid having its base horizontal and its cusp on the axis of x: these conditions as to the position of the cycloid must be understood in all that follows with respect to the present problem, although for the sake of brevity we may omit to state them explicitly.

Hence the curve we require can consist of nothing except a portion or portions of such a cycloid, together with a portion or portions of the circular arc. See Art. 18.

It is obvious that the solution cannot consist of a cycloid alone, as if there were no condition imposed: for such a cycloid would have a cusp at A, and therefore its tangent vertical there; and so it would be initially *outside* the circle.

150. We therefore proceed to investigate whether the necessary conditions will be satisfied by the curve composed of a circular portion AR, a cycloidal portion RS, and a circular portion SB; or of two out of these three possible portions; or of one only.

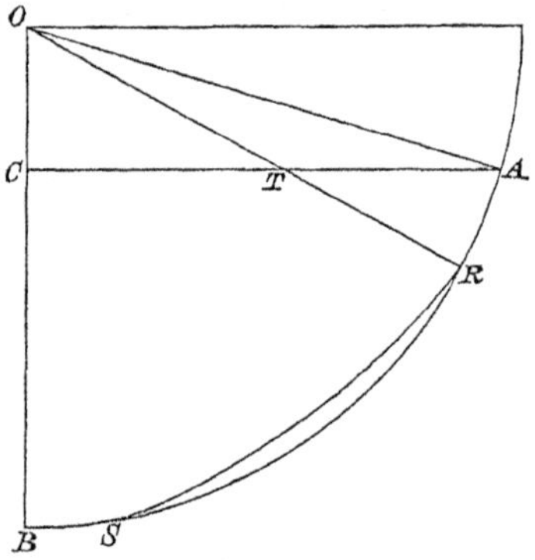

Let O be the centre of the circle; draw AC horizontal: let $OA = r$, $OC = b$.

The equation to the circle is

$$(x - c)^2 + (y + b)^2 = r^2,$$

c being a constant depending on the point in AC which we take as origin.

For the circle we have

$$N = -\frac{r}{2y^{\frac{3}{2}}(y + b)}, \qquad P = -\frac{x - c}{ry^{\frac{1}{2}}};$$

and so it will be found that

$$N - \frac{dP}{dx} = \frac{y - b}{2ry^{\frac{3}{2}}}.$$

For the cycloid we know that

$$N - \frac{dP}{dx} = 0.$$

Our object is to ensure that δu shall be positive. The value of δy corresponding to any part of the circular arc is *necessarily negative;* so that $\int\left(N - \frac{dP}{dx}\right) \delta y \, dx$ is necessarily positive for limits corresponding to the circular arc provided $y - b$ is negative.

151. Now we observe in the first place that the circle and the cycloid must touch at R, supposing that R does not coincide with A.

For let P_0 be the value of P corresponding to the circle at R, and let P_1 be the value of P corresponding to the cycloid at R. Then the part of δu which is free from the integral sign gives rise at the point R to the term $(P_0 - P_1)\, \delta y$. Now δy is necessarily negative at R; and therefore $P_0 - P_1$ must be zero or negative: it cannot be negative, for then the cycloid would fall outside the circle. Hence $P_0 - P_1$ must be zero; and therefore the circle and the cycloid must touch at R.

Of course if R coincide with A we have only $-P_1 \,\delta y$ for the corresponding part of δu, and this vanishes since δy vanishes: thus it is not necessary that the circle and the cycloid should touch.

In like manner if S does not coincide with B the circle and the cycloid must touch at S.

152. We shall now consider what results follow from supposing the circle and the cycloid to touch at R. See the diagram to Art. 150.

Let OR and AC intersect at T; let the angle OTC be denoted by θ, and the angle OAC by α. Thus

$$RT = r - \frac{b}{\sin\theta} = r - \frac{r\sin\alpha}{\sin\theta}.$$

Now from the properties of the cycloid we know that the diameter of the generating circle is $\frac{RT}{\sin\theta}$, that is $\frac{r(\sin\theta - \sin\alpha)}{\sin^2\theta}$.

If the cycloid touches the circle again at S we obtain another expression for the diameter of the generating circle by ascribing to θ the value which it has at S instead of the value which it has at

R: and of course these two expressions must be equal in value. This leads us to examine the values of which the expression $\frac{\sin\theta - \sin\alpha}{\sin^2\theta}$ is susceptible.

Denote it by v; then

$$\frac{dv}{d\theta} = \frac{\cos\theta\,(2\sin\alpha - \sin\theta)}{\sin^3\theta}.$$

If 2 $\sin\alpha$ is less than unity the maximum value of v is given by $\sin\theta = 2\sin\alpha$; and then v may *twice* have an assigned value, namely, once when θ is less than the value which corresponds to the maximum, and once when θ is greater than this value.

If $2\sin\alpha$ is not less than unity then the maximum value of θ is given by $\cos\theta = 0$; and as θ changes from α to $\frac{\pi}{2}$ the value of v continually increases. In this case then we do not get the same value of v for two admissible values of θ; and it is impossible to draw an arc of a cycloid RS which touches the circle at two points. Nor is it possible to draw an arc of a cycloid which touches the circle at a point R and passes through B. For the diameter of the generating circle would be here less than the maximum value of $\frac{r(\sin\theta - \sin\alpha)}{\sin^2\theta}$, that is less than $r(1 - \sin\alpha)$, that is less than CB, and so the cycloid could not pass through the point B.

Hence we arrive at the result that the circular arc itself is the line of quickest descent. It will be observed that $y - b$ is equal to $r(\sin\theta - 2\sin\alpha)$, and this is in the present case negative for the whole of the circular arc; and so $\int\left(N - \frac{dP}{dx}\right)\delta y\,dx$ is necessarily positive for admissible values of δy. Here there is no discontinuity.

153. Let us now return to the case in which $2\sin\alpha$ is less than unity. In this case at B and adjacent to B we have $y - b$ positive for the circle, and so $\int\left(N - \frac{dP}{dx}\right)\delta y\,dx$ would be negative for admissible values of δy. Hence the arc of the circle adjacent

to B cannot form a portion of the required line of quickest descent. Thus the arc of the cycloid must pass through B.

Hence finally, if $2\sin\alpha$ is less than unity the required curve consists of a portion AR of the circle, and then a cycloidal arc touching the circle at R and passing through B; that is, we have the discontinuity of the arcs of two different curves which touch at the common point. And the common point R must be such that here y is less than b; and in fact this is ensured by the nature of the cycloid. For as the cycloid is to touch the circle internally the radius of curvature of the cycloid must be less than the radius of the circle; that is twice TR must be less than OR; therefore TR is less than TO, which makes y less than b for the point R.

It is obvious that $2\sin\alpha$ is greater or less than unity according as the circular arc BA is less or greater than an arc of $60°$.

154. It may be useful to shew that it is possible to draw such a cycloid as we have supposed when $2\sin\alpha$ is less than unity.

Suppose we take $r-b$ for the diameter of the generating circle of the cycloid, and put the vertex of the cycloid at B; this cycloid will fall *without* the circle at B, because $2r-2b$, which is the radius of curvature of the cycloid at the vertex, is by supposition greater than r. On the other hand, if the diameter of the generating circle of the cycloid is indefinitely great, and the cycloid be made to pass through B, it will obviously fall entirely within the circle. Starting from the last case diminish the diameter of the generating circle of the cycloid continuously, making the cycloid always pass through B. Then we must arrive at the case in which the cycloid just touches the circle before cutting it. The point of contact will not be at A, for the tangent to the cycloid would then be vertical, while the tangent to the circle would not be vertical. The point of contact will not be at B, for there the tangent to the circle is horizontal, while the tangent to the cycloid would not be horizontal. Hence the contact must take place at some intermediate point, as we require.

155. Required the curve of quickest descent from a fixed point A to a fixed point B, with the condition that the path is not

to pass inside the given circular arc AB, which does not exceed a quadrant, B being the lowest point of the circle.

After the discussion of the problem enunciated in Art. 149 it will be sufficient to state the results of the present problem.

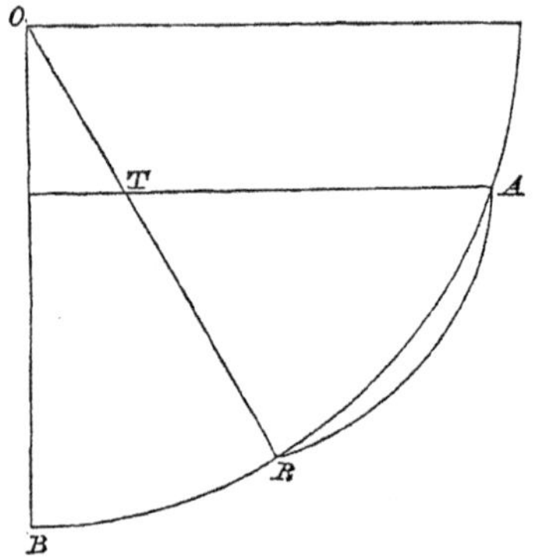

The required curve consists of the arc of a cycloid having its cusp at A and touching the circle at some point R, and the portion RB of the circle. As the cycloid is outside the circle at R twice TR is greater than OR, and therefore TR is greater than OT. Hence for the point R we have y greater than b, and thus $\int \frac{y-b}{2ry^{\frac{3}{2}}} \delta y \, dx$ is necessarily positive for the part RB of the path, since δy is positive.

If AB is an arc of 90° there is no cycloidal portion, and the required curve consists entirely of the circular quadrant.

156. The problems of Arts. 149 and 155 include as special cases a problem proposed by the late Dr Whewell in the Smith Prize Papers for 1846. His enunciation was as follows: Prove that an arc of a circle from the lowest point which does not exceed 60° is a curve of quicker descent than any other curve which can be drawn *within* the same arc; and that the arc of 90° is a curve of quicker descent than any other curve which can be drawn *without* the same arc.

It would be interesting to know how the distinguished philosopher treated the problem himself. It will be seen that in our investigation we obtain for the circular arc

$$\delta u = \int \frac{(y-b)\,\delta y\,dx}{2ry^{\frac{3}{2}}};$$

this expression shews at once that the results enunciated are true with respect to such curves as differ *infinitesimally* from the given circular arc; but it would still remain to shew that the results are true for curves which differ to a *finite extent* from the given circular arc. The investigation which we have supplied establishes the results completely, including them in fact in wider statements.

157. Some extension may be given to the problem of Art. 149.

Required the curve of quickest descent from a fixed point A to a fixed point B, with the condition that the path is not to pass

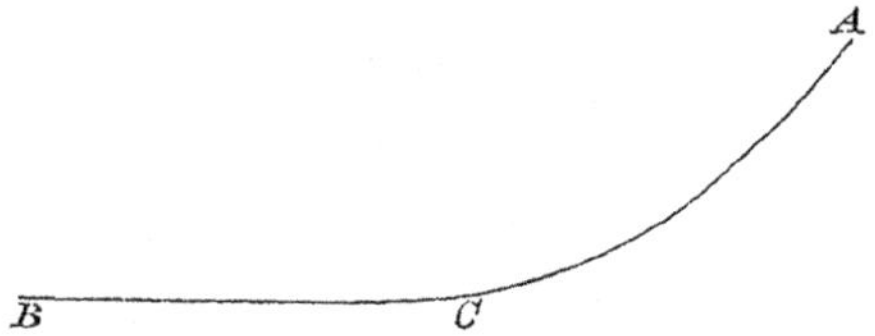

outside a certain boundary ACB composed of a circular arc AC, not exceeding a quadrant, and the straight line CB, which is the tangent to the arc at C, the lowest point of the circle.

If AC is not greater than 60°, the boundary is the required curve.

If AC is greater than 60°, the required curve consists of a part of AC beginning at A; then an arc of a cycloid which touches AC at the point of departure, and either passes through B or touches BC: if the arc of the cycloid touches BC, the last portion of the required curve is a portion of BC. It depends on the length of BC whether the cycloid passes through B or touches BC.

It will be seen that, taking u as in Art. 149, we have for the part of the boundary which coincides with BC

$$\delta u = -\int \frac{\delta y\, dx}{2y^{\frac{3}{2}}};$$

this is positive, since δy is here necessarily negative.

158. Some extension may also be given to the problem of Art. 155.

Required the curve of quickest descent from a fixed point A to a fixed point B, with the condition that the path is not to pass

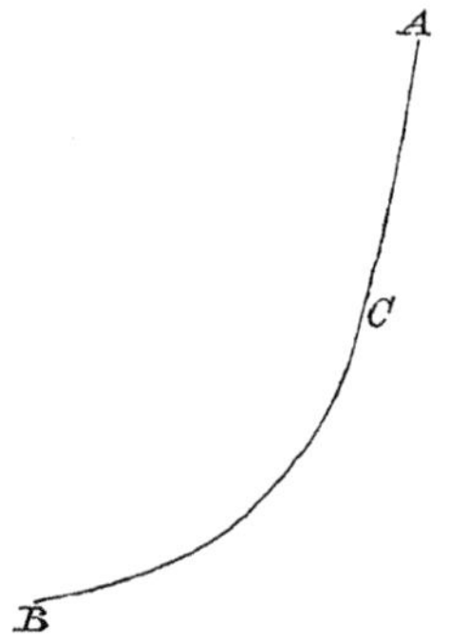

inside a certain boundary composed of a circular arc BC, not exceeding a quadrant, B being the lowest point of the circle, and the straight line CA which is a tangent to the circle at C.

The required curve consists in general of an arc of a cycloid having its cusp at A, and touching BC; and of the arc of the circle from the point of contact to B. If BC is a quadrant, however, the boundary itself is the required curve.

The expression $N - \frac{dP}{dx}$ of Art. 149 reduces for the straight line AC to $-\frac{1}{2y^{\frac{3}{2}}\sqrt{1+p^2}}$; and as δy is here essentially positive, the integral $-\int \frac{\delta y\, dx}{2y^{\frac{3}{2}}\sqrt{1+p^2}}$ is negative: thus AC cannot form part of the required curve. If, however, AC is *vertical* we have p infinite, so that the integral just written may be considered to vanish: this agrees with our statement that AC is now part of the required curve. But it would be prudent in this case to choose polar co-ordinates, so as to avoid the infinite value of p.

Take as the initial line the horizontal line through A; then we have to find a minimum value of

$$\int \frac{\left\{r^2 + \left(\frac{dr}{d\theta}\right)^2\right\}^{\frac{1}{2}} d\theta}{(r \sin\theta)^{\frac{1}{2}}}.$$

Call this u; then

$$\delta u = P\delta r + \int \left(N - \frac{dP}{d\theta}\right) \delta r \, d\theta,$$

where
$$N = \frac{r}{(r\sin\theta)^{\frac{1}{2}} \left\{r^2 + \left(\frac{dr}{d\theta}\right)^2\right\}^{\frac{1}{2}}} - \frac{\left\{r^2 + \left(\frac{dr}{d\theta}\right)^2\right\}^{\frac{1}{2}}}{2r^{\frac{3}{2}} (\sin\theta)^{\frac{1}{2}}},$$

$$P = \frac{\frac{dr}{d\theta}}{(r\sin\theta)^{\frac{1}{2}} \left\{r^2 + \left(\frac{dr}{d\theta}\right)^2\right\}^{\frac{1}{2}}}.$$

Now take for the equation to the vertical straight line

$$r\cos\theta = c;$$

then
$$\frac{dr}{d\theta} = \frac{c\sin\theta}{\cos^2\theta} = r\tan\theta; \quad \left\{r^2 + \left(\frac{dr}{d\theta}\right)^2\right\}^{\frac{1}{2}} = \frac{r}{\cos\theta}.$$

Hence we shall find that

$$N = \frac{\cos\theta}{(r\sin\theta)^{\frac{1}{2}}} - \frac{1}{2\cos\theta\,(r\sin\theta)^{\frac{1}{2}}},$$

$$P = \left(\frac{\sin\theta}{r}\right)^{\frac{1}{2}};$$

and from these values it will follow that

$$N - \frac{dP}{d\theta} = 0.$$

Thus the part of δu which corresponds to the vertical straight line vanishes.

159. Examples involving discontinuity connected with the brachistochrone might easily be multiplied: we will give another.

Required the line of quickest descent from the fixed point A to the fixed point B, under the condition that the tangent to the path shall not make with the horizon an angle greater than a given angle α.

The following will be the solution in general:

AC is a straight line inclined to the horizon at an angle α;

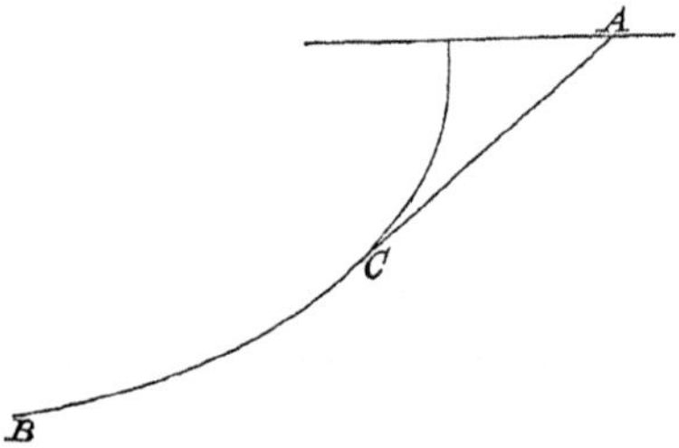

and CB is an arc of a cycloid which has its cusp in the horizontal line through A, and touches AC at C.

The part of δu which corresponds to AC will be as in Art. 158,

$$-\int \frac{\delta y\, dx}{2y^{\frac{3}{2}}\sqrt{1+p^2}}, \text{ that is } -\frac{\cos\alpha}{2}\int \frac{\delta y\, dx}{y^{\frac{3}{2}}};$$

and this is positive, because δy is necessarily negative.

It might happen that B is on the ascending part of the cycloid, and that if it were too high up it would be necessary to have a rectilinear portion at the end of the path as well as at the beginning.

If the straight line joining A and B is inclined to the horizon at an angle α, this straight line is the solution.

Suppose it also required that the tangent to the path shall not make with the horizon an angle less than β. Then, if the tangent to BC is never inclined to the horizon at an angle less than β, the solution already given still holds. But if at any point between B and C the tangent is inclined to the horizon at an angle less than β, this solution does not hold. In this case we must draw through B a straight line inclined to the horizon at an angle β; then find a cycloid having its cusp in the horizontal line through A which will touch both the straight lines, namely, that through A and that through B. The required curve consists of three portions: the straight line from A to the point of contact with the cycloid; the arc of the cycloid; and the straight line from the point of contact to B.

160. The last example which we will consider is the following:

Required the curve of quickest descent from the fixed point A to the fixed point B, under the condition that the radius of curvature of the path shall never be less than a given constant r, and that there shall be no abrupt change of direction.

By the principles of the Calculus of Variations every thing is excluded from the boundary except a cycloid with its base horizontal and its cusp in the horizontal line through A, and a circle of radius r. For if any other curve occurred in the boundary there would be an existent variation of the first order which could be made to have either sign, and so could be made negative. The circle is not excluded because on account of the condition respecting the radius of curvature it may be impossible to give either sign to δy. Moreover, the required curve cannot *begin* with the cycloid at A; for the radius of curvature of the cycloid at its cusp is indefinitely small.

161. In some cases a solution will be furnished by the use of a circle of radius r passing through A and B. For suppose such a circle drawn, and suppose that the height of its centre above the horizontal line through A is not less than the depth of B below this horizontal line. Then we have as in Art. 150,

$$\delta u = \int \frac{(y-b)\,\delta y\,dx}{2ry^{\frac{3}{2}}};$$

and $y-b$ is never positive throughout the arc. Thus if δy is negative, δu is necessarily positive. And δy cannot be positive, for then the condition that the radius of curvature is never to be less than

r would be somewhere broken. Hence if the given radius r be large enough we have one solution in the arc of a circle. It is easy to state a limiting value of r. Draw a horizontal straight line as far above A as B is vertically below A. Bisect AB at right angles by a straight line which meets the horizontal straight line just drawn at a point which we will call S. Then if r is not less than AS this solution will hold; but it will not hold if r is less than AS.

There must then be another solution which will hold at least in some cases.

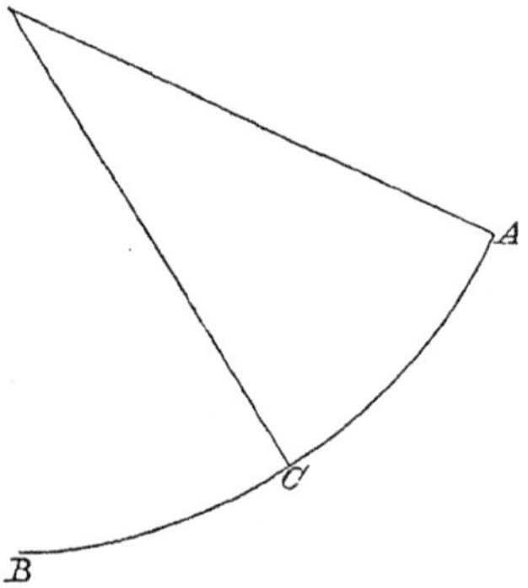

Let AC be an arc of a circle of radius r; and CB be an arc of a cycloid with its base horizontal and a cusp in the horizontal line through A, touching the circle at C and having its radius of curvature at C not less than r. The length of AC and the position of its centre, and the parameter of the cycloid must be determined so as to satisfy these conditions, and to make the time down ACB a minimum: this constitutes an ordinary problem of the Differential Calculus, and we may assume that it can be theoretically solved. This gives us the line which we require.

To justify this solution we have only to proceed in the manner of Arts. 85 and 119. For a special kind of variation in the path the value of δu to the first order is zero; this special kind of variation being that produced by passing from AC to an arc of radius r infinitesimally above, or infinitesimally below AC. We know that

δu to the first order is here zero, because by hypothesis we have obtained a minimum in our Differential Calculus problem. And for any other admissible kind of variation we shall find that δu to the first order is positive. After the example discussed in Art. 85 it will not be necessary to be very elaborate here. I will take only one case.

Let E be the point in AC where $y-b$ changes sign. Let the curve AFG be one obtained by an admissible variation. Draw

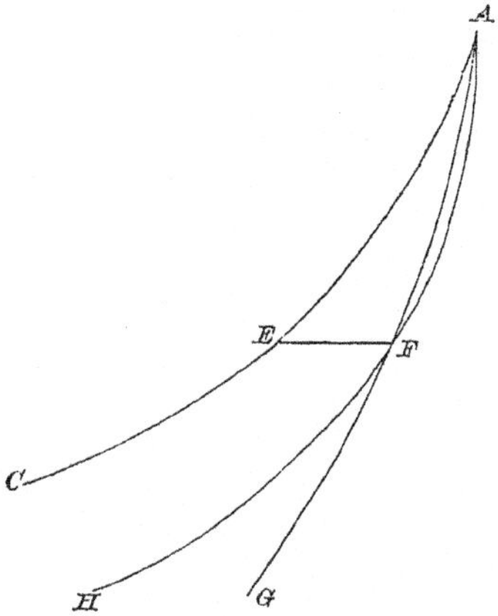

EF horizontal. Let AFH be an arc of circle of radius r. Since along the curve AFG the radius of curvature is not to be less than r, the curve and the circular arcs must be situated as in the diagram.

Now corresponding to AFH and the cycloid from H the value of δu to the first order is zero, as we have seen. In passing from AFH to the curve AFG there is a diminution of negative matter corresponding to the area between the two arcs from A to F; and there is a gain of positive matter corresponding to the area between FG and FH.

We have hitherto tacitly supposed that in the cycloidal part CB the radius of curvature is never less than r; but it may happen that in some cases this condition cannot be fulfilled: then our solution must be composed of *three* parts, the first and the last being arcs of circle of radius r, and the intermediate part an arc of a cycloid.

162. The discussions in Arts. 147...161 illustrate completely the general principles we have laid down. In every case there is a discontinuity arising from a condition or conditions imposed. If we vary over the whole extent of our proposed solutions we find that the variation of the time is a small positive quantity of the first order, which may vanish in a particular case. If we vary over the cycloidal portion alone the variation of the time becomes an indefinitely small quantity of the second order. But we did not attempt to examine this term; for it is obvious that there must be some path of *least* time in all the problems we have examined, and that the Calculus of Variations rejects every solution except that which we have given in each case respectively.

CHAPTER VIII.

PROBLEM OF LEAST ACTION.

163. LET $u = \int \psi(x)\,\phi(p)\,dx$; and suppose it be required to find the maximum or minimum value of u, the limiting values of the variables being fixed.

Here we obtain in the usual way to the second order

$$\delta u = \int \psi(x)\,\phi'(p)\,\delta p\,dx + \frac{1}{2}\int \psi(x)\,\phi''(p)\,(\delta p)^2\,dx$$

$$= \psi(x)\,\phi'(p)\,\delta y - \int \frac{d}{dx}\{\psi(x)\,\phi'(p)\}\,\delta y\,dx$$

$$+ \frac{1}{2}\int \psi(x)\,\phi''(p)\,(\delta p)^2\,dx.$$

The term of the first order which is outside the integral sign vanishes, since the limits are fixed. Then as usual we must have

$$\psi(x)\,\phi'(p) = C, \text{ a constant.}$$

Thus δu reduces to

$$\frac{1}{2}\int \psi(x)\,\phi''(p)\,(\delta p)^2\,dx.$$

Consider the equation

$$\psi(x)\,\phi'(p) = C \quad\quad (1);$$

this gives p in terms of x, from which y must be obtained. There will thus be two constants which must be determined by making the curve pass through the given extreme points.

It is possible that in some cases a solution would be found by taking $C=0$ and $\phi'(p)=0$; the solution would consist of one or more straight lines, corresponding to the various values of p. This case would resemble that of Art. 6.

Suppose that when p is infinite $\phi'(p)$ has a finite value b; then

$$\psi(x)=\frac{C}{b} \quad \text{.........................} \quad (2)$$

is a singular solution or a particular solution of (1). For (2) gives constant values for x; and a constant value of x corresponds to $p=$ infinity. In this case however on account of the infinite value of p it would be prudent to give another investigation, by polar co-ordinates for instance.

However it appears that we may expect cases of discontinuity; and we will now take a particular example.

164. Let $u=\int\sqrt{(x+a)(1+p^2)}\,dx$. Then (1) becomes

$$\frac{p\sqrt{(x+a)}}{\sqrt{(1+p^2)}}=\sqrt{c} \quad \text{.......................} \quad (3),$$

where for symmetry we put $\sqrt{c}$ instead of C.

Thus
$$p^2=\frac{c}{x+a-c};$$

therefore
$$y-C'=\pm 2\sqrt{c(x+a-c)} \quad \text{................} \quad (4),$$

where C' is a constant.

And as $\frac{p}{\sqrt{(1+p^2)}}$ is unity, when p is infinite we have

$$x+a=c$$

for a singular solution of (3).

Suppose the origin is taken at one of the fixed points; then from (4)

$$-C'=\pm 2\sqrt{c(a-c)}.$$

Thus (4) becomes

$$y \pm 2\sqrt{c(a-c)} = \pm 2\sqrt{c(x+a-c)} \quad \text{..........(5).}$$

This equation represents two parabolas if $c(a-c)$ is positive.

Let (h, k) denote the other fixed point through which the required curve is to pass: then

$$k \pm 2\sqrt{c(a-c)} = \pm 2\sqrt{c(h+a-c)},$$

so that $$k^2 \pm 4k\sqrt{c(a-c)} = 4ch;$$

therefore $$(k^2 - 4ch)^2 = 16k^2 c(a-c) \quad \text{..........(6);}$$

therefore $$16c^2(h^2+k^2) - 8c(h+2a)k^2 + k^4 = 0 \quad \text{..........(7).}$$

From (7) we have two real values of c provided

$$(h+2a)^2 \text{ is greater than } k^2 + h^2,$$

that is provided k^2 is less than $4a(a+h)$.

It is plain from (6) that these values of c make $c(a-c)$ positive.

Hence we obtain two parabolas provided that k^2 is less than $4a(a+h)$.

165. The example we are now discussing may be enunciated thus: determine the path of a particle for which the *action* $\int v\,ds$ is a minimum between fixed points, if the velocity v at any point is that due to a fall from the straight line $x + a = 0$; the axis of x being vertically downwards.

If a particle is projected from a point with a given velocity, and acted on by gravity, we know that it will describe a parabola; and if the particle is to pass through a second fixed point so long as this point lies within a certain boundary there are two parabolas either of which may be taken. Hence we conclude that our two parabolas are the two which might be described by a particle under the action of gravity if it started with the proper velocity.

The equation $k^2 = 4a(a+h)$ determines the boundary within which the second fixed point must be situated in order that the projectile starting with the proper velocity may reach it; this boundary is a parabola; we shall call it the *bounding parabola.*

The value of δu, by Art. 163, is now

$$\frac{1}{2}\int\frac{\sqrt{(x+a)}}{(1+p^2)^{\frac{3}{2}}}(\delta p)^2\,dx.$$

If x increases algebraically constantly throughout the integration this expression is certainly positive. If x diminishes algebraically during part of the integration, then during that part we have to take $(1+p^2)^{\frac{1}{2}}$ and $(1+p^2)^{\frac{3}{2}}$ as negative; so that the above expression is still positive. If then the arc of the parabola which we have to consider does not include the vertex, we may feel certain that we have a minimum. But if it does include the vertex, since at that point p is *infinite,* we cannot be certain that we have a minimum: the change of sign of $(1+p^2)^{\frac{1}{2}}$ at the vertex also would render our process suspicious.

166. Of course when the second fixed point is beyond the bounding parabola neither of the two parabolic paths is possible: so that some other minimum path *must* exist for this case, and *may* exist even if one or both of the parabolas is also a minimum. We have then to discover this other solution; and also to settle the doubtful point noticed in the preceding Article as to whether a parabolic path is a minimum.

167. We shall first investigate whether a solution can be obtained by combining an arc of a parabola with the straight line defined by $x+a=c$: for we have already seen in Art. 164 that the last equation furnishes a singular solution of the differential equation (3) of that Article.

It is essential that c in (3) should retain the same value at a point of discontinuity in order that the part of δu which is outside the sign of integration may vanish. Hence p must not undergo any abrupt change at a point of discontinuity. Thus if such a solution as we are now examining can exist, it must consist of a parabolic arc and the tangent *at the vertex* produced.

Thus we have two cases to consider; the parabolic arc being described before or after the straight line according as the point to be reached is higher or lower than the starting point.

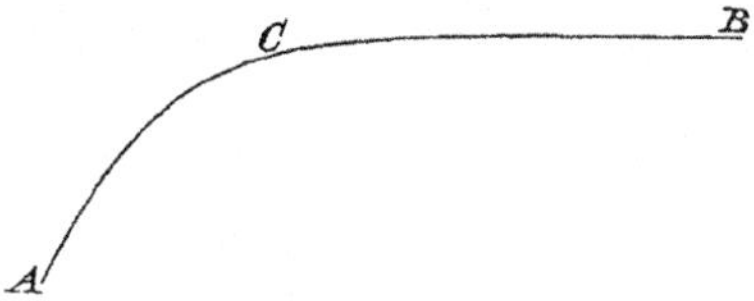

A is the starting point, B the other fixed point, C the vertex of the parabola.

We have to determine whether this solution really gives a minimum.

If we take the axis of x vertically downwards, as hitherto, the value of p is infinite at C; and so δp may be very great in the neighbourhood of C, and we cannot depend on the investigation of Art. 163. Let us then take the axis of y vertically downwards. Thus

$$u = \int \sqrt{(y+a)(1+p^2)}\, dx;$$

$$\delta u = \int \left\{ \frac{\sqrt{(1+p^2)}}{2\sqrt{(y+a)}} \delta y + \frac{p\sqrt{(y+a)}}{\sqrt{(1+p^2)}} \delta p \right\} dx.$$

Thus it is obvious that $p = 0$ gives neither a maximum nor a minimum; for it leaves us with

$$\delta u = \int \frac{\delta y\, dx}{2\sqrt{(y+a)}} \quad\ldots\ldots\ldots\ldots\ldots\ldots\ldots\ldots (8),$$

and we can of course make this positive or negative as we please.

We have then still to seek a path of minimum action for the case in which the path to be reached is beyond the bounding parabola.

168. Now we have already in the course of our investigations seen that a solution of a problem in the Calculus of Variations may be furnished, at least in part, by some bounding line which by the

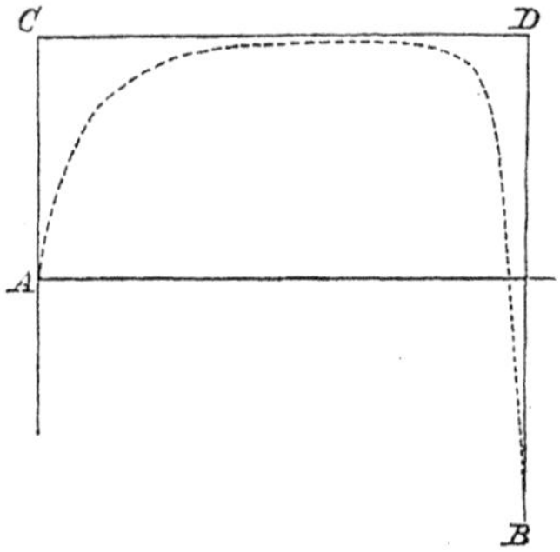

circumstances of the case cannot be transgressed: and we shall find that such is the case with the present problem.

Let A be the starting point, B the point to be reached; let CD be a horizontal straight line at a vertical distance a above A.

Then CD is such a bounding line as we have supposed; for above CD the expression for the velocity becomes impossible. We shall shew presently that the integral $\int v ds$ is a minimum for the path composed of the straight lines AC, CD and DB. Strictly speaking we cannot have a particle describing this path, because the velocity is zero at any point of CD. But we may suppose a curve drawn very close to this path; and the action along the curve will only differ infinitesimally from the value of the integral taken over the path.

169. We shall now shew that the integral $\int v ds$ really is a minimum for the path composed of AC, CD and DB.

This will require some other investigation than those hitherto given in this Chapter in order to avoid infinite quantities. The equation (8) of Art. 167, for instance, is now unsatisfactory because $y+a$ is zero along CD.

Suppose Ac on the curve equal in length to AC on the straight line. Divide AC and Ac into the same number of equal infinitesimal elements. Let PQ and pq be a corresponding pair; so that

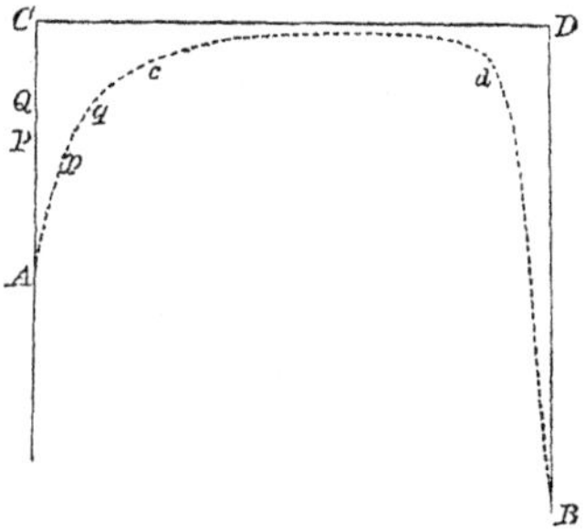

Ap is equal in length to AP. Then P will be vertically higher than p, and so the velocity at P less than the velocity at p; and therefore the action over PQ will be less than the action over pq. In this way we see that the whole action over AC is less than the whole action over Ac. Similarly if Bd is equal in length to BD the whole action over DB is less than the whole action over Db.

And finally the action along CD is zero, and that along cd is not zero.

Thus the action for the path made up of AC, CD and DB is certainly less than the action for the adjacent curve.

170. We will now return to the point which was left unsettled in Art. 165, and determine under what circumstances a parabolic arc is really a minimum.

We will take the axis of y vertically downwards, and put the origin at the highest possible point; that is at a point where the velocity must be zero. Thus we have

$$u = \int \sqrt{y(1+p^2)}\, dx.$$

Hence proceeding in the ordinary way to find the minimum, we see that we must have

$$y = c_1 + \frac{(x - c_2)^2}{4c_1} \quad\text{.......................}(9),$$

where c_1 and c_2 are constants.

Then by Jacobi's theory, in Art. 23, the term of the second order in δu is

$$\frac{1}{2}\int \frac{\sqrt{y}}{(1+p^2)^{\frac{3}{2}}} \left\{\frac{1}{z}\frac{dz}{dx}\delta y - \delta p\right\}^2 dx,$$

where $$z = b_1 \frac{dy}{dc_1} + b_2 \frac{dy}{dc_2},$$

b_1 and b_2 being arbitrary constants.

Thus $$z = b_1 \left\{1 - \frac{(x - c_2)^2}{4c_1^{\,2}}\right\} - b_2 \frac{x - c_2}{2c_1}$$

$$= b_1 \left\{1 - \frac{b_2}{b_1}\omega - \omega^2\right\},$$

where ω stands for $\dfrac{x - c_2}{2c_1}$.

We see from (9) that $\omega = p$.

Now we know it is essential for a maximum or a minimum that z should not vanish during the range of the integration. To make z vanish we require that

$$\frac{b_2}{b} = \frac{1}{p} - p.$$

If then $p - \dfrac{1}{p}$ can take every value between $-\infty$ and $+\infty$ we are sure that z will vanish during the range of the integration. If $\dfrac{1}{p} - p$ does not take every value between $-\infty$ and $+\infty$ we can select $\dfrac{b_2}{b_1}$ so that z shall not vanish.

If the arc of the parabola extends from one end of any focal chord to the other end $\frac{1}{p} - p$ ranges from $-\infty$ to ∞, and there is not a minimum. If the arc of the parabola is of greater extent, then of course there is not a minimum. If the arc is of less extent there is a minimum. That is, to ensure a minimum it is necessary and sufficient that the arc should subtend at the focus an angle less than two right angles.

171. Suppose that a projectile is to start from A with a given velocity, and to pass through B. Then we know that the focus of the path must be in the intersection of two circles, one described

from the centre A, and the other from the centre B. Thus one parabola will have its focus above AB, and the other will have its focus below AB. The former parabola does not give a minimum action, the latter does.

This is in harmony with a result we shall presently obtain, namely, that of two parabolic paths between the same points the action is always greater for the higher parabola than for the lower. And the example furnishes an excellent illustration of Jacobi's process. If an arc of a parabola be greater than would be cut off by any focal chord, suppose a portion of the arc cut off by a chord passing below the focus but very close to the focus. Then suppose this arc removed and replaced by another with the same chord, but having its focus as much below the chord as the focus of the original arc is above the chord. Thus we can get a path differing only infinitesimally from the former, but giving a less action than the former: therefore the former arc cannot be an arc of minimum action.

172. Hence our final conclusion is the following: the discontinuous path composed of three straight lines is always a path of minimum action; and for points outside the bounding parabola, as there is no other path of minimum action, this must be the path of *least* action. For points inside the bounding parabola there is also another path of minimum action, namely, the lower of the two parabolas which could be described by a projectile between the two points. Hence for points inside the bounding parabola, as there are two, and only two, paths of minimum action, one of them will be the path of *least* action. As we shall see presently sometimes the continuous path is that of *least* action, and sometimes the discontinuous path; one or the other being necessarily such.

173. We will now estimate the value of the action in various cases.

In a parabolic path the velocity being that due to a fall from the directrix is $\sqrt{2rg}$, where r represents the radius vector from the focus. Suppose $4n$ the latus rectum; then $r = \dfrac{n}{\cos^2 \frac{\theta}{2}}$; therefore $\dfrac{ds}{d\theta} = \dfrac{n}{\cos^3 \frac{\theta}{2}}$.

Thus the action is $\sqrt{2gn^3} \displaystyle\int \frac{d\theta}{\cos^4 \frac{\theta}{2}}$.

Now we know that if a particle were to describe a parabola under a force *in the focus* of the absolute intensity μ the *time* of describing an arc is $\dfrac{n^{\frac{3}{2}}}{\sqrt{2\mu}} \displaystyle\int \frac{d\theta}{\cos^4 \frac{\theta}{2}}$.

Thus we see that the action in the parabolic arc with which we are concerned is equal to $2\sqrt{\mu g}$ into the time of describing the same parabolic arc under a force in the focus.

Hence we are able to express the *action* we require by the aid of Lambert's theorem respecting the *time* in a parabolic arc.

Suppose r_1 and r_2 the radii vectores of the extremities of the arc, and c the chord which joins these extremities. Then for the arc which subtends less than two right angles at the focus the action is

$$\frac{\sqrt{g}}{3}\{(r_1+r_2+c)^{\frac{3}{2}}-(r_1+r_2-c)^{\frac{3}{2}}\};$$

and for the arc which subtends more than two right angles at the focus the action is

$$\frac{\sqrt{g}}{3}\{(r_1+r_2+c)^{\frac{3}{2}}+(r_1+r_2-c)^{\frac{3}{2}}\}.$$

Thus we see that the action is greater for the latter arc than for the former.

By the discontinuous path the action is that which arises in going vertically upwards through a space r_1 and downwards through a space r_2; therefore the action is

$$\frac{2\sqrt{2g}}{3}\{(r_1^{\frac{3}{2}}+r_2^{\frac{3}{2}}\}.$$

Thus in any particular case we might determine by calculation whether the parabola of minimum action or the discontinuous locus gives the *least* action.

For example, suppose the second fixed point to be *on the bounding parabola;* then the two parabolic paths coincide so that the common focus is on the straight line AB; therefore we have $c=r_1+r_2$: thus the action for the parabolic path is

$$\frac{2\sqrt{2g}}{3}(r_1+r_2)^{\frac{3}{2}};$$

this is greater than the action for the discontinuous path, for $(r_1+r_2)^{\frac{3}{2}}$ is greater than $r_1^{\frac{3}{2}}+r_2^{\frac{3}{2}}$.

174. The quantity denoted by r_1 in the preceding Article is the same as we had previously denoted by a. If we keep to the *same* parabola, but change the position of the second fixed point by gradually increasing r_2 we shall arrive at a point such that the action in the discontinuous path is less than the action in the minimum parabola; for the parabola will touch the bounding parabola at some point, and to reach the point of contact we know that the discontinuous solution gives the least action.

175. It is easy to see by a diagram that if the minimum parabolic path to any point gives a greater action than the corresponding discontinuous path, then this inequality also holds for any *further* point on the *same* parabola.

The diagram shews that if the action through AB is greater than the action through AP, PQ, QB; then *a fortiori* the action through Ab is greater than the action through AP, Pq, qb.

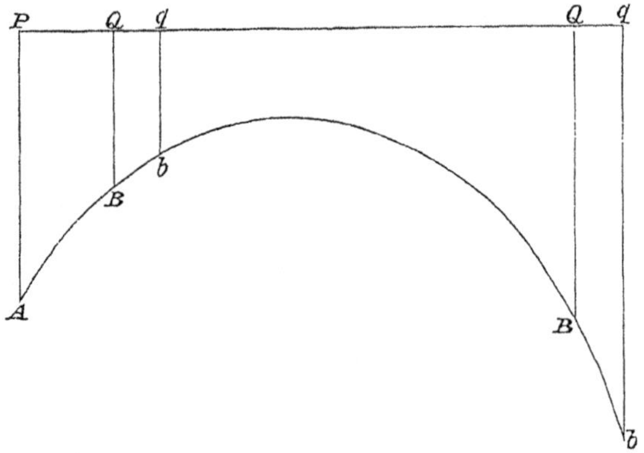

Moreover, when B is near enough to A the action through AB is certainly less than the action through AP, PQ, QB.

176. Thus we may conclude that for points within a certain boundary, the action is least for the minimum parabolic path, and that for points beyond this boundary the action is least for the discontinuous path. The boundary is determined by the condition

$$(r_1 + r_2 + c)^{\frac{3}{2}} - (r_1 + r_2 - c)^{\frac{3}{2}} = 2\sqrt{2}\,(r_1^{\frac{3}{2}} + r_2^{\frac{3}{2}}).$$

This condition may be easily converted into an ordinary rectangular or polar equation. For instance, to convert it into a polar equation, we put

$$a \text{ for } r_1, \quad r \text{ for } c, \quad a - r\cos\theta \text{ for } r_2;$$

thus the equation becomes

$$\left(a + r\sin^2\frac{\theta}{2}\right)^{\frac{3}{2}} - \left(a - r\cos^2\frac{\theta}{2}\right)^{\frac{3}{2}} = a^{\frac{3}{2}} + (a - r\cos\theta)^{\frac{3}{2}}.$$

[Hence, when $\theta = 0$ we have

$$a^{\frac{3}{2}} - (a - r)^{\frac{3}{2}} = a^{\frac{3}{2}} + (a - r)^{\frac{3}{2}};$$

therefore $r = a$.

When $\theta = \frac{\pi}{2}$ we have

$$\left(a + \frac{r}{2}\right)^{\frac{3}{2}} - \left(a - \frac{r}{2}\right)^{\frac{3}{2}} = 2a^{\frac{3}{2}}.$$

This gives $r = a\sqrt{\sqrt{192} - 12} = a \times 1{\cdot}36$ nearly.

When $\theta = \pi$, we have

$$(a + r)^{\frac{3}{2}} - a^{\frac{3}{2}} = a^{\frac{3}{2}} + (a + r)^{\frac{3}{2}};$$

this requires r to be infinite.]

177. The problem of the present Chapter has been discussed in the *Quarterly Journal of Mathematics* for November, 1868; but I have added considerably to the discussion which is there given. The discontinuity which presents itself is very similar to that which we had in Chapter IV.; the reason of the discontinuity here being the condition which exists that the required curve cannot go above a certain horizontal line.

178. We will now modify the preceding problem slightly: Suppose that a heavy particle is to move in a smooth tube from one fixed point to another, starting with a given velocity: required the path so that the action may be the least possible with the condition that the path is never to go higher than the higher of the two fixed points.

It is obvious that there must be some path of least action. By the Calculus of Variations everything is excluded from the path except a horizontal straight line and a parabola with its axis vertical. Hence if an arc of a parabola alone will not satisfy the conditions, we shall be constrained to adopt the solution given by the diagrams of Art. 167. If we ascribe any variation to the figures, the variation of the action to the first order will be zero for the parabolic part and positive for the rectilinear part:

the term of the second order in the variation of the action is necessarily positive as in Art. 170. Hence we are certain that we have a minimum; and this minimum must give us the *least* action, as there is no other minimum.

The position of C in the diagrams is easily determined; for the vertical height of C is known; and thus the latus rectum of the parabola is known; and so we can find the horizontal distance between C and A or B.

179. The investigations of the present Chapter naturally suggest for discussion the following problem: a particle is projected from a given point with a given velocity, and is attracted to a fixed point by a force varying inversely as the square of the distance: determine the path of minimum action to a second fixed point.

Take the straight line from the first fixed point to the centre of force as the initial line; let μ denote the intensity of the force, and r_1 the distance of the starting point from the centre of force. It will be sufficient for our purpose to limit ourselves to one of the three cases which might exist, and suppose that the velocity of projection is less than $\sqrt{\frac{2\mu}{r_1}}$. Denote the velocity of projection by $\sqrt{\left(\frac{2\mu}{r_1}-\frac{\mu}{a}\right)}$. Then we know that at any distance r from the centre of force the velocity will be

$$\sqrt{\left(\frac{2\mu}{r}-\frac{\mu}{a}\right)}.$$

Put p for $\frac{dr}{d\theta}$, and let

$$u=\int\sqrt{\left(\frac{2}{r}-\frac{1}{a}\right)}\sqrt{(r^2+p^2)}\,d\theta,$$

then we require the minimum value of u.

By the usual theory of the subject, we obtain

$$\left(\frac{2}{r}-\frac{1}{a}\right)^{\frac{1}{2}}\frac{r^2}{\sqrt{(r^2+p^2)}}=\text{a constant};\ \ldots\ldots\ldots\ldots (1).$$

The integral of this is known to be

$$r = \frac{a(1-e^2)}{1+e\cos(\theta-\beta)} \quad\text{.....................} \quad (2),$$

where e and β are constants; for this reduces the value of the left-hand member to the constant $\sqrt{a(1-e^2)}$. This integral may easily be obtained by finding $\frac{d\theta}{dr}$ from the differential equation, and changing the variable from r to the reciprocal of r. For denoting the constant by $\sqrt{b}$ we get

$$p^2 = \left(\frac{2}{r} - \frac{1}{a}\right)\frac{r^4}{b} - r^2.$$

Put $\frac{1}{u}$ for r: thus

$$\left(\frac{du}{d\theta}\right)^2 = \left(2u - \frac{1}{a}\right)\frac{1}{b} - u^2;$$

therefore by differentiation

$$\frac{d^2u}{d\theta^2} + u = \frac{1}{b}.$$

This is immediately integrable.

Now (2) is the equation to an ellipse, since the constant e is less than unity; this ellipse has its focus at the centre of force, and has $2a$ for its major axis.

180. As the ellipse is to pass through a second given point, we have now to determine if such an ellipse can always be drawn so as to fulfil all the conditions stated.

Since the major axis is $2a$, the distance of the second focus from the starting point is $2a - r_1$; so that this focus lies on a circle described from the starting point as centre with the radius $2a - r_1$. In like manner this focus of the required ellipse lies on a circle described from the second fixed point as centre with radius $2a - r_2$ where r_2 is the distance of the second fixed point from the centre of force. If these two circles intersect, there are two positions for the second focus, and two ellipses can be obtained; if the circles touch, one ellipse can be obtained; if the circles do not intersect, there is no such ellipse as we require.

181. The boundary of the space within which the second fixed point must be situated in order that an elliptic arc may connect the two fixed points is an ellipse, as we shall now shew.

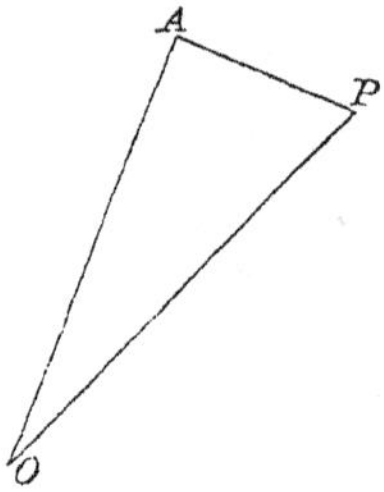

Let O be the centre of force, A the fixed starting point, P any point on the boundary: then we must have

$$AP = 2a - OA + 2a - OP;$$

therefore $$OP + AP = 4a - OA.$$

Thus the locus of P is an ellipse of which O and A are the foci, and the major axis is $4a - OA$: the excentricity is the ratio of OA to the major axis, that is $\frac{r_1}{4a - r_1}$.

182. Now guided by the analogy of our preceding investigations, we may anticipate the following results for our problem of minimum action.

Denote the fixed points by A and B. There is always a discontinuous minimum solution which is thus obtained: the radii vectores to A and to B must be produced until they meet the circle described with O as centre and the radius $2a$; the produced parts of the radii vectores together with the arc of the circle which they intercept constitute the solution. It may be shewn to be a true minimum by the process of Art. 169. If the second fixed point is outside the bounding ellipse this is the only minimum, and therefore gives the path of least action.

If the second fixed point is inside the bounding ellipse, two elliptic arcs can be drawn between the two fixed points; only one of these, however, is a curve of minimum action, namely, that which has both its foci on the same side of the straight line which joins the two fixed points: this can be shewn by Jacobi's method as will presently appear. This example of his method was indicated by Jacobi himself. [See Todhunter's *History of the Calculus of Variations*, page 251.]

Thus, if the second fixed point is inside the bounding ellipse, there are two paths of minimum action, one continuous and the other discontinuous; in some cases one of the two is the path of *least* action, and in some cases the other is: one of the two being necessarily the path of least action.

183. We will now determine by Jacobi's method which of the two elliptic arcs gives a path of minimum action. We denote by O the centre of force; then O being the origin the equation to an ellipse is

$$r = \frac{a(1-e^2)}{1+e\cos(\theta-\beta)}.$$

Let S denote the other focus. It is obvious that for one ellipse S and O are on the same side of AB, and for the other ellipse S and O are on opposite sides of AB; in the former case the arc with which we are concerned subtends at S an angle less than two right angles, and in the latter case an angle greater than two right angles. We shall shew that in the former case the path is one of minimum action, but not in the latter case.

The quantity denoted by z in Art. 25 here stands for

$$b_1 \frac{dr}{d\beta} + b_2 \frac{dr}{de},$$

where b_1 and b_2 are arbitrary constants.

Now $$\frac{dr}{d\beta} = -e\frac{dr}{d\theta} = -ep.$$

To find $\frac{dr}{de}$ we have

$$\frac{a(1-e^2)}{r} = 1 + e\cos(\theta - \beta);$$

therefore
$$-\frac{2ae}{r} - \frac{a(1-e^2)}{r^2}\frac{dr}{de} = \cos(\theta-\beta)$$
$$= \frac{1}{e}\left\{\frac{a(1-e^2)}{r} - 1\right\};$$

hence
$$\frac{dr}{de} = \frac{r^2 - ar(1+e^2)}{ae(1-e^2)}.$$

Thus
$$z = C_1\left\{r^2 - ar(1+e^2) + mp\right\},$$

where C_1 and m are arbitrary constants.

Hence we must examine the range of values of which

$$\frac{r^2 - ar(1+e^2)}{p}$$

is susceptible. We shall find that if the elliptic arc subtends at S an angle of two right angles, the expression is susceptible of all values between $-\infty$ and $+\infty$; and of course this will therefore be true if the angle is greater than two right angles: hence in these cases there will not be a minimum. But if the elliptic arc subtends at S an angle less than two right angles the expression is not susceptible of all values, and so there will be a minimum.

It is obvious that $\frac{r^2 - ar(1+e^2)}{p}$ is infinite and changes sign at a vertex of the ellipse; hence we shall establish our point if we shew that at the two ends of a chord through the focus this expression has the same value and the same sign; for then the expression, as it begins with a certain value, changes sign as it passes through infinity, and finally returns to its original value, will have passed through all values.

Now
$$\frac{r^2 - ar(1+e^2)}{p} = \frac{1 - \frac{a}{r}(1+e^2)}{\frac{p}{r^2}}$$

$$= \frac{1 - \frac{a}{r}(1+e^2)}{\frac{e \sin(\theta - \beta)}{a(1-e^2)}};$$

the value varies as

$$\frac{r - a(1+e^2)}{r \sin(\theta - \beta)}.$$

Put $2a - \rho$ for r, and $\rho \sin \phi$ for $r \sin(\theta - \beta)$; so that ρ is the radius vector to the focus S, and ϕ is the angle which this radius vector makes with the major axis of the ellipse. Hence the expression becomes

$$\frac{a(1-e^2) - \rho}{\rho \sin \phi}.$$

Now $\sin \phi$ has numerically the same value at the two ends of the chord through S, but with opposite signs; let ρ_1 and ρ_2 denote the two values of ρ. Then we have only to shew that

$$\frac{a}{\rho_1}(1-e^2) - 1 = -\left\{\frac{a(1-e^2)}{\rho_2} - 1\right\},$$

that is
$$a(1-e^2)\left(\frac{1}{\rho_1} + \frac{1}{\rho_2}\right) = 2;$$

and this is obviously true.

184. The *action* through any arc of the ellipse described under a force at O is known to vary as the *time* of describing the same arc under a force at S; the theorem of Art. 173 is indeed a particular case of this. Thus we might express the action as in Art. 173 by using the theorem for an ellipse, which is analogous to Lambert's theorem for a parabola.

185. It is easy to illustrate discontinuous solutions still further, by an example like that of Art. 178. Required a path

of minimum action between the two fixed points A and B with the condition that the path shall lie entirely between two circles, one described with centre O and radius OA, the other described with centre O and radius OB. It is possible that an arc of an ellipse alone will satisfy the conditions. If not, the path must be composed of an arc of an ellipse having one extremity at one fixed point, and touched at its vertex by an arc of a circle described from O as centre with a radius equal to the distance from O of the other fixed point.

[This will be sufficiently obvious from what has been already frequently stated. The path can consist of nothing but some combination of an arc of an ellipse with an arc of the prescribed boundary. It is necessary that the two arcs should *touch* at the common point in order that the term in the variation which is free from the integral sign may vanish. Thus the common point must be a vertex of the ellipse.

Let u be taken as in Art. 179. Then it will be found that for the arc of the circular boundary δu reduces to $\int \frac{(a-r)\,\delta r}{\sqrt{(2a-r)\,ra}}\,d\theta$. This is positive. For suppose that the arc of the circle is part of the *larger* circle; then δr is necessarily negative along the circular arc. And $a-r$ is negative. For, by hypothesis, the elliptic solution which would be applicable if there were no condition imposed is now inapplicable; that is, this ellipse would cross the boundary. But by Art. 183 this ellipse has both its foci on the same side of AB: and this makes the distance from O of the more remote of the two points A and B greater than a; that is, r is greater than a.

In like manner if the arc of the circle is part of the *smaller* circle δu is positive; for then δr and $a-r$ are both positive.]

CHAPTER IX.

SOLIDS OF MINIMUM RESISTANCE.

186. To find the form of a solid of revolution which experiences a minimum resistance when it moves through a fluid in the direction of the axis of revolution.

I need scarcely say that the interest of this problem arises from its historical connexion with many illustrious names, including that of Newton; and is thus quite independent of the amount of practical value of the results, and of the trustworthiness of the ordinary theory of the resistance of fluids.

Take the axis of x as that of revolution; then by the ordinary principles we require a minimum of

$$\int \frac{yp^3}{1+p^2}\,dx.$$

Hence, in the usual way, we obtain

$$\frac{yp^3}{1+p^2} = yp\,\frac{3p^2(1+p^2) - 2p^4}{(1+p^2)^2} + \text{constant};$$

therefore $$-\frac{2yp^3}{(1+p^2)^2} = \text{constant} \quad\text{................}\quad (1).$$

Denote the constant by $-2c_1$; then we have

$$y = \frac{c_1(1+p^2)^2}{p^3} \quad\text{........................}\quad (2).$$

Hence, by differentiation,

$$p = \frac{c_1 (p^2+1)(p^2-3)}{p^4} p \frac{dp}{dy}.$$

Thus $\frac{dp}{dy}$ is positive or negative according as p^2 is greater or less than 3; for we may take c_1 to be positive: in the former case the generating curve is convex to the axis of x, and in the latter case concave.

Now denoting $\frac{p^3}{1+p^2}$ by $\phi(p)$, we find that

$$\phi'(p) = \frac{3p^2+p^4}{(1+p^2)^2},$$

and
$$\phi''(p) = \frac{2p(3-p^2)}{(1+p^2)^3}.$$

Hence by the general investigation given in Art. 26 we have for the term of the second order in δu

$$\int \frac{(3-p^2)p}{(1+p^2)^3} \left\{ y(\delta p)^2 - y''(\delta y)^2 \right\} dx.$$

It follows at once that there will be no minimum if p^2 is greater than 3 within the range of the integration. But if p^2 is less than 3 throughout this range the curve is concave to the axis of x as we have already seen: thus y'' is negative, and we have necessarily a minimum.

And
$$x = \int \frac{dy}{p} = c_1 \int \frac{(p^2+1)(p^2-3)}{p^5} dp$$

$$= c_1 \left\{ \log p + \frac{1}{p^2} + \frac{3}{4p^4} \right\} + c_2 \quad \text{..................} \quad (3),$$

where c_2 is an arbitrary constant.

Let us now suppose that the generating curve is to terminate at fixed points; so that the surface is in fact to be a zone of a surface of revolution. From (2) and (3) theoretically we must eliminate p, and thus obtain an equation between x and y and the constants c_1 and c_2. Then having given the co-ordinates of the two fixed points we have two equations for determining the constants. Thus theoretically all is satisfactory. But there are some important remarks to make on the solution.

187. We must be careful not to speak of our solution as giving the solid of *least* resistance. Legendre in fact pointed out that by

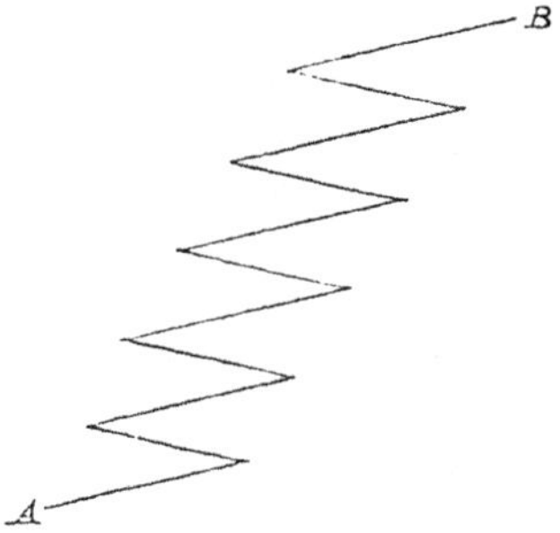

taking a zigzag line for our generating curve we might make the resistance as small as we please. [See Todhunter's *History of the Calculus of Variations,* page 229.]

Our solution gives us only a solid of minimum resistance; by which we mean that if we pass from our generating curve to another by giving to y and p *infinitesimal* changes, we shall obtain a solid of greater resistance. But our investigation does not allow us to pass from our curve joining the fixed points A and B to the zigzag line; because although the change in y might be infinitesimal throughout, yet the change in p would not be such.

It should be observed that Legendre's zigzag line may be supposed to be suggested by the fundamental equation (1) of Art. 186; inasmuch as $p = 0$ is *a* solution of that equation, namely, when the constant is zero. Legendre himself does not make this remark, which is however important; for otherwise we might be left with a feeling of general distrust, and the expectation that solutions will present themselves in an arbitrary manner without the warning of the fundamental equation.

188. Suppose however we add to our problem the condition that the generating curve is to have p always with the *same sign*. It is obvious that there must be some curve with this condition which generates the solid of *least* resistance; and our investigation assures us that it can be no other than the curve which we have obtained; unless it be some discontinuous line to which we shall presently advert. Hence finally; if we do not impose the condition that p is to be of invariable sign, we can only say that our solution gives a minimum with respect to *infinitesimal* variations of y and p; but if we impose the condition, we may assert that our solution gives the solid of *least* resistance unless any discontinuous solution can be found.

189. But it is easy to see that the continuous solution which we have obtained cannot be universally applicable.

Let A and B be the fixed points; and suppose that the straight line drawn from A to B is inclined to the axis of x at an angle

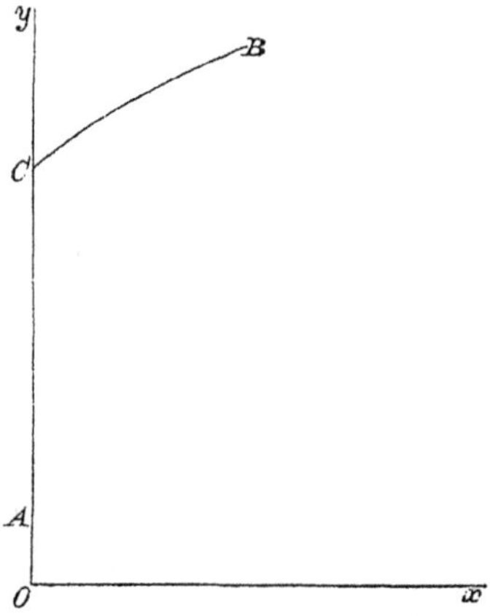

greater than 60°: then it is certain that we cannot draw a curve from A to B with the condition that p^2 shall always be less than 3. We must therefore seek for another solution.

Now if we limit p to have always the same sign, we do in effect determine that our curve shall not fall outside the rectangle which has AB for a diagonal, and has two of its sides parallel to the direction of motion.

Therefore we naturally proceed to enquire if the boundary thus assigned does not itself in part constitute the generating curve.

Suppose the axis of y to pass through A. We propose to seek for a solution composed of AC, which is part of this axis, and a curve CB which satisfies equation (2).

190. Let $OC = y_0$, and $OA = k$; let the abscissa of B be a, and the ordinate b.

Then we seek for a minimum of

$$\pi (y_0^2 - k^2) + 2\pi \int_0^a \frac{yp^3 dx}{1+p^2}.$$

Put u for $y_0^2 + 2\int_0^a \frac{yp^3 dx}{1+p^2}$.

Then find δu, and make the part of δu which is of the first order and under the integral sign vanish: when this is done, the remaining term of the first order in δu is

$$2y_0 \left\{1 - \frac{3p^2 + p^4}{(1+p^2)^2}\right\}_0 \delta y_0,$$

the subscript 0 indicating that the values correspond to the point C. To make this term vanish, we must have

$$\left\{1 - \frac{3p^2 + p^4}{(1+p^2)^2}\right\}_0 = 0,$$

which leads to $p_0 = 1$. Thus the curve must meet the axis of y at an angle of 45°.

Now consider the part of δu which is of the second order. From the term y_0^2 in u there arises $(\delta y_0)^2$. From the other term in u, remembering that y_0 is not constant, we have by the general investigation of Art. 26,

$$-\{\phi'(p)\,(\delta y)^2\}_0 + \int_0^a \phi''(p)\,\{y\,(\delta p)^2 - y''(\delta y)^2\}\,dx,$$

where $\phi(p)$ stands for $\frac{p^3}{1+p^2}$.

Thus we have $\phi'(p)=1$ when $p=1$. Hence δu reduces to

$$2\int_0^a \frac{(3-p^2)\,p}{(1+p^2)^3}\,\{y\,(\delta p)^2 - y''(\delta y)^2\}\,dx,$$

which is essentially positive.

191. It only remains to enquire if real values can be found for the constants c_1 and c_2 which occur in our discontinuous solution.

Let ϖ denote the value of p at B; then since $p=1$ at A, we have, by equations (2) and (3),

$$y_0 = 4c_1,$$

$$0 = c_1\left(1+\frac{3}{4}\right)+c_2,$$

$$b = \frac{c_1(1+\varpi^2)^2}{\varpi^3},$$

$$a = c_1\left(\log\varpi+\frac{1}{\varpi^2}+\frac{3}{4\varpi^4}\right)+c_2\,;$$

from these we have to determine c_0, c_1, y_0 and ϖ.

Substitute the value of c_2 from the second of these equations in the last; then we have

$$y_0 = 4c_1 \quad\text{..}\quad (4),$$

$$b = \frac{c_1(1+\varpi^2)^2}{\varpi^3} \quad\text{..................................}\quad (5),$$

$$a = c_1\left\{\log\varpi+\frac{1}{\varpi^2}+\frac{3}{4\varpi^4}-\frac{7}{4}\right\} \quad\text{......................}\quad (6).$$

From (5) and (6) we determine ϖ; for by division we have

$$\frac{b}{a} = \frac{(1+\varpi^2)^2}{\varpi^3\log\varpi+\varpi+\dfrac{3}{4\varpi}-\dfrac{7}{4}\varpi^3}\,;$$

since the expression on the second side of this equation would change from infinity to zero, as ϖ changed from 1 to 0, a real value of ϖ to satisfy the equation must exist. Then from (5) we find c_1, and from (4) we find y_0.

192. In order, however, that this solution may hold, we must have y_0 greater than k, that is $4c_1$ greater than k. It is not in our power to give a simple expression for the relation which must hold between a, b, and k at the limit of the possibility of the solution. We can however shew that if $b-k$ is greater than a the possibility is ensured.

For from (5) and (6) we have

$$b - a = c_1 z \quad \text{.........................} \quad (7),$$

where z stands for

$$\frac{(1+\varpi^2)^2}{\varpi^3} - \left(\log\varpi + \frac{1}{\varpi^2} + \frac{3}{4\varpi^4} - \frac{7}{4}\right);$$

thus $$\frac{dz}{d\varpi} = \frac{(1-\varpi)(3+2\varpi^2-\varpi^4)}{\varpi^5}.$$

Hence we see that $z=4$ when $\varpi=1$, and that z diminishes as ϖ diminishes from the value unity.

It follows then from (7) that $4c_1$ is never less than $b-a$; and as $b-a$ is by hypothesis greater than k, we have $4c_1$ greater than k.

Thus the solution which we have obtained is certainly admissible in some cases in which the continuous solution is not admissible, namely, the cases in which $b-k$ is greater than $a\sqrt{3}$.

193. The next question is whether any other solution can be found. The ordinary theory of the Calculus of Variations assures us that there can be no *continuous* solution except that which we first obtained, or, in other words, that there is no other solution in which δy is susceptible of either sign. Thus there can be no other solution than the continuous solution, and a solution or solutions composed in part of the boundary to which we are by supposition restricted. We have already investigated one such discontinuous solution. Two other cases present themselves for examination. We may try if the generating line can be composed in part of a straight line parallel to the axis of revolution drawn through the more remote of the two fixed points. But

it may be ascertained immediately that no solution of this kind exists. For the upper limit of integration will now not be fixed; denote this limit by x_1. Then we should require a minimum of

$$y_0^2 + 2\int_0^{x_1} \frac{yp^3}{1+p^2}\,dx;$$

and hence in the variation there arises a term

$$2\left(\frac{yp^3}{1+p^2}\right)_1 dx_1,$$

which does not vanish.

The only remaining case for examination is that in which the generating line consists of the curve AC, and the portion CB of the ordinate at B.

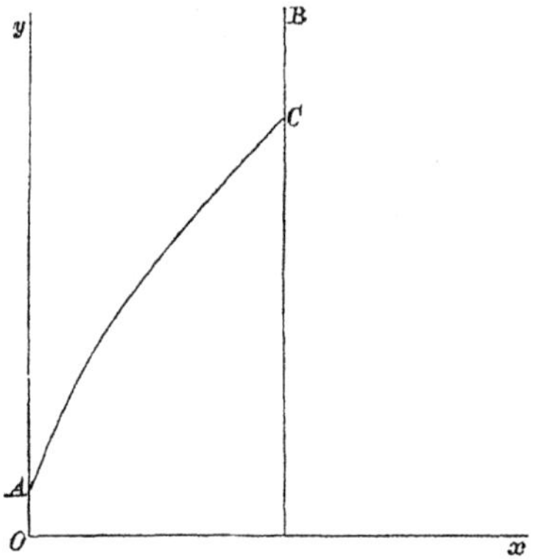

Let y_1 denote the ordinate of C. Proceed as in the former case. We now seek a minimum of

$$-y_1^2 + 2\int_0^a \frac{yp^3}{1+p^2}\,dx.$$

The term of the first order in the variation outside the integral sign will now be

$$2y_1\left\{\frac{3p^2+p^4}{(1+p^2)^2} - 1\right\}_1 \delta y_1.$$

To make this vanish we must have $p_1 = 1$, so that the curve must now meet the ordinate of B at an angle of 45°.

The term of the second order in the variation arising from $-y_1^2$ is cancelled by a similar term arising from the variation, of $2\int_0^a \frac{yp^3}{1+p^2}dx$; and we are finally left with the same essentially positive value of the term of the second order in the variation as we had in Art. 190.

194. To determine the constants c_1 and c_2 we have the following relations, where ϖ now denotes the value of p at A.

$$0 = c_1\left(\log\varpi + \frac{1}{\varpi^2} + \frac{3}{4\varpi^4}\right) + c_2,$$

$$k = \frac{c_1(1+\varpi^2)^2}{\varpi^3},$$

$$a = c_1\left(1 + \frac{3}{4}\right) + c_2,$$

$$y_1 = 4c_1.$$

And ϖ must lie between 1 and $\sqrt{3}$; also $4c_1$ must be less than b.

By eliminating c_2 and c_1 we arrive at the following equation:

$$\frac{a}{k} = \frac{\varpi^3}{(1+\varpi^2)^2}\left\{\frac{7}{4} - \log\varpi - \frac{1}{\varpi^2} - \frac{3}{4\varpi^4}\right\}.$$

As ϖ varies from 1 to $\sqrt{3}$ the right-hand member of this expression varies from 0 to $\frac{3\sqrt{3}}{16}\left(\frac{4}{3} - \frac{\log 3}{2}\right)$, so that if $\frac{a}{k}$ lies between these limits a real value of ϖ can be found to satisfy the equation. But this solution only gives a solid of minimum resistance, and not the solid of least resistance, as will appear from the next Article.

195. In Art. 190 we see that the tangent to the curvilinear part of the solution never makes with the axis of revolution an angle greater than 45°. This result is in harmony with an interesting proposition given by the late R. L. Ellis; see *Quarterly Journal of Mathematics*, Vol. x. p. 122. It follows from Mr Ellis's proposition that our solids would not be solids of least resistance

even with the condition of having p always positive if the value of p were ever greater than unity. The proposition of R. L. Ellis is a generalization of one given by Newton: see the *Principia*, Book II. Prop. 34. Newton may be said to be the first person who treated a problem of the Calculus of Variations; and that problem involved a discontinuous solution.

196. Let us now finally sum up the results we have obtained with respect to the continuous solution and the two discontinuous solutions.

By virtue of Legendre's remark already noticed none of these solutions gives us a solid of *least* resistance.

Every one of these solutions when it really exists gives us *a* solid of *minimum* resistance; that is, any admissible variation in the form of the generating curve would increase the resistance; by an admissible variation, we mean that δy and δp must be always infinitesimal, and that our curve is to be comprised between the extreme ordinates, so that no variation is required for x.

If we impose the condition that p is to be of invariable sign, then in every case one of our solutions gives *the* solid of least resistance; namely, *the* solution which exists, supposing that only one exists, and that for which the resistance is least when more than one exists.

197. The discontinuity which occurs in the preceding investigations may be considered to arise from the conditions which we impose, namely, explicitly that p is to be of invariable sign, and implicitly that the curve is to be comprised between the extreme ordinates. The fundamental relation of Art. 186, namely

$$-\frac{2yp^3}{(1+p^2)^2} = \text{constant},$$

may be considered to be satisfied in a certain sense by $p = \infty$, and this may suggest a straight line parallel to the axis of y as forming a part of the solution. But unless the conditions be imposed we shall not be able to shew that we have really a minimum.

For example, if in Art. 189 we were at liberty to extend to the left of the axis of y we might replace AC by a boundary which would give a less resistance.

198. A particular case of the preceding problem may deserve special notice, namely, that in which one of the fixed points is on the axis of revolution.

Generally suppose we require a maximum or minimum of $\int y\phi(p)\,dx$. By the ordinary method we obtain

$$y\{\phi(p) - p\phi'(p)\} = \text{constant}.$$

If one of the extreme points is on the axis of x, and $\phi(p) - p\phi'(p)$ is never infinite, the constant must be zero.

In the present case we thus obtain

$$\frac{yp^3}{(1+p^2)^2} = 0.$$

This will not furnish us with any *minimum*, except we regard as such Legendre's zigzag line which may be supposed to correspond to $p = 0$. We may consider that we obtain a *maximum* by combining $y = 0$ with $p = \infty$, that is, by taking a portion of the axis of x and the ordinate of the second fixed point; for thus we obviously obtain the *greatest* value.

If we impose the condition that p is to be always of one sign, we shall obtain a minimum like that of Arts. 189...191; the solution is always applicable, for we shall not now require as in Art. 192 that $4c_1$ shall be greater than an assigned positive quantity. And as no other minimum presents itself, we infer that we have the figure of least resistance, under the assigned condition.

199. The following elementary problem will serve to illustrate the result which we obtained in the discontinuous solution of Art. 190, that the curve meets the initial ordinate at an angle of 45°.

A and B are fixed points, P is a variable point on the ordinate at A: it is required to determine the position of P, so that the

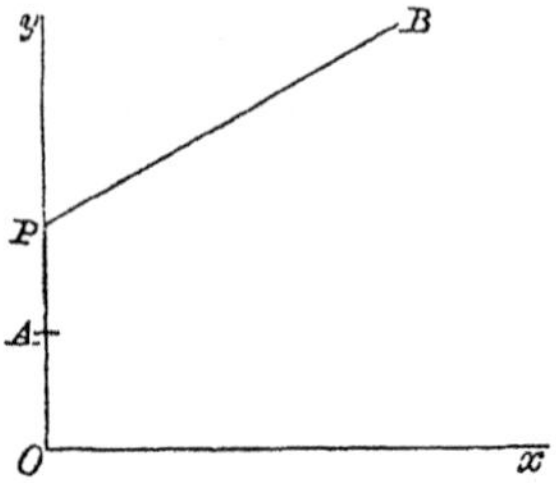

resistance on the surface generated by the revolution of the straight lines AP and PB may be the least possible.

Let $OP = y$; let $OA = k$; let b be the ordinate of B and a the abscissa. Let θ be the inclination of PB to the axis of revolution; then $b - y = a \tan \theta$.

The resistance varies as $y^2 - k^2 + (b^2 - y^2) \sin^2 \theta$, that is as $y^2 \cos^2 \theta + b^2 \sin^2 \theta - k^2$, that is as $(b \cos \theta - a \sin \theta)^2 + b^2 \sin^2 \theta - k^2$, that is as $b^2 - k^2 - 2ab \sin \theta \cos \theta + a^2 \sin^2 \theta$, that is as

$$b^2 - k^2 + \frac{a^2}{2} - ab \sin 2\theta - \frac{a^2}{2} \cos 2\theta,$$

that is as
$$b^2 - k^2 + \frac{a^2}{2} - \frac{a\sqrt{(4b^2 + a^2)}}{2} \cos (2\theta - \alpha),$$

where
$$\cos \alpha = \frac{a}{\sqrt{(4b^2 + a^2)}}.$$

The resistance is therefore least when $2\theta = \alpha$.

If a be very small this gives very nearly $2\theta = \frac{\pi}{2}$.

We may observe that if θ is greater than $\frac{\pi}{4}$ our expression for the resistance increases with θ.

200. Another elementary investigation may also be usefully supplied here.

Suppose PQ an indefinitely small arc of a curve; PR and RQ any indefinitely small arbitrary straight lines: required an

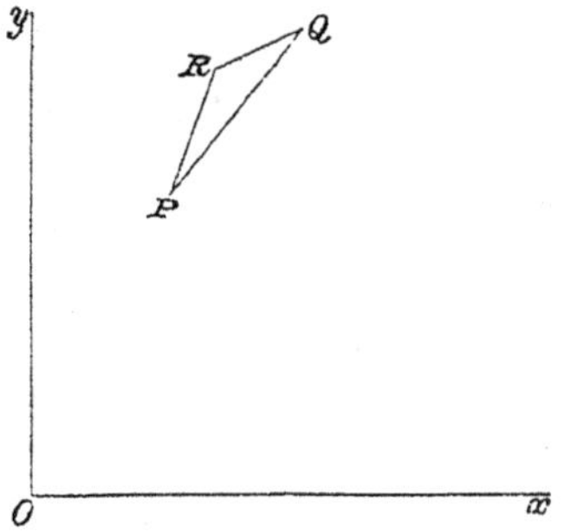

expression for the difference between the resistance on the strip of surface generated by the revolution of PQ and the sum of the resistances on the strips of surface generated by the revolution of PR and RQ.

Let x and y be the co-ordinates of P; $x + dx$ and $y + dy$ the co-ordinates of Q. Let $PR = v$, $RQ = w$. Let α be the inclination of PR to the axis of x, and β the inclination of RQ to the axis of x. Let $PQ = ds$.

Now omitting a certain factor in the usual notation, which is constant, the resistance on the strip of surface generated by the revolution of PQ will be ultimately $2y\left(\frac{dy}{ds}\right)^2 dy$; the resistance on the strip of surface generated by the revolution of PR will be ultimately $2yv\sin^3\alpha$; and the resistance on the strip of surface generated by the revolution of RQ will be ultimately $2yw\sin^3\beta$.

Hence we require the value of

$$2y\left\{\left(\frac{dy}{ds}\right)^2 dy - v\sin^3\alpha - w\sin^3\beta\right\}.$$

But $$v \sin \alpha + w \sin \beta = dy,$$

and $$v \cos \alpha + w \cos \beta = dx;$$

therefore $$v = \frac{dy \cos \beta - dx \sin \beta}{\sin (\alpha - \beta)},$$

and $$w = \frac{dx \sin \alpha - dy \cos \alpha}{\sin (\alpha - \beta)}.$$

Thus our expression becomes

$$2y \left\{ \left(\frac{dy}{ds}\right)^2 dy - \frac{(dy \cos \beta - dx \sin \beta) \sin^3 \alpha + (dx \sin \alpha - dy \cos \alpha) \sin^3 \beta}{\sin (\alpha - \beta)} \right\};$$

and this $= 2y \left\{ \left(\frac{dy}{ds}\right)^2 dy + dx \sin \alpha \sin \beta \sin (\alpha + \beta) \right.$

$$\left. - \frac{dy}{\sin (\alpha - \beta)} [\cos \beta \sin \alpha (1 - \cos^2 \alpha) - \cos \alpha \sin \beta (1 - \cos^2 \beta)] \right\}$$

$$= 2y \left\{ \left(\frac{dy}{ds}\right)^2 dy + dx \sin \alpha \sin \beta \sin (\alpha + \beta) \right.$$

$$\left. - dy + dy \cos \alpha \cos \beta \cos (\alpha + \beta) \right\}.$$

By putting this into factors we obtain

$$\frac{2y}{ds^2} (dy \cos \alpha - dx \sin \alpha) (dy \cos \beta - dx \sin \beta) (dy \cos \gamma + dx \sin \gamma);$$

where γ stands for $\alpha + \beta$.

For a particular case, suppose $\alpha = \frac{\pi}{2}$ and $\beta = \frac{\pi}{4}$; then our expression becomes

$$\frac{y\, dx\, (dy - dx)^2}{ds^2};$$

and this is positive.

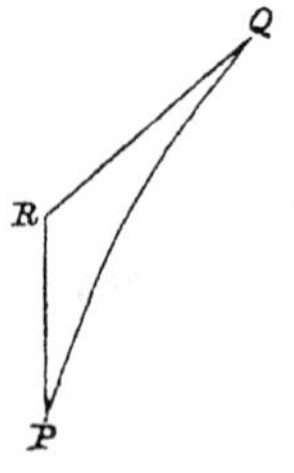

This particular case and the next particular case are given in the *Quarterly Journal of Mathematics*, Vol. X. page 122.

This shews that if we have a curve such that the tangent is everywhere inclined to the axis of x at an angle greater than

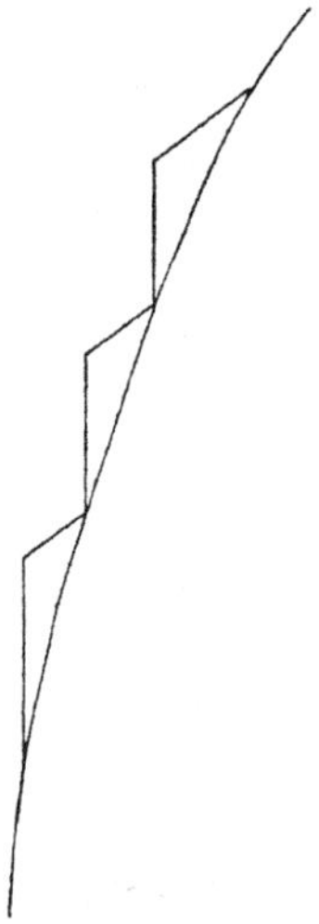

45°, we can diminish the resistance on the solid generated, by passing from the curve to the notched boundary where the straight lines are alternately inclined at 90° and 45° to the axis.

For another particular case of the general result, let PR be inclined to the axis of x at 45° and RQ at 90°; then we shall

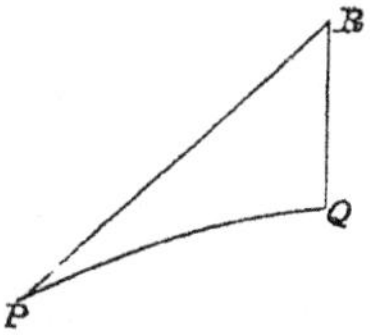

find that the sum of the resistances corresponding to PQ and QR exceeds the resistance corresponding to PR by

$$\frac{y\,dx\,(dy - dx)^2}{ds^2};$$

and this is positive.

201. Now let AP be any arc of a curve, such that the tangent PT at P is inclined to the axis of x at 45^0; let AT be inclined to the axis of x at 90^0. We proceed to compare the resistance on the surface generated by the revolution of AP with the resistance on the surface generated by the revolution of AT and TP.

Let $OA = y_0$, $OT = a$; let h and k be the co-ordinates of P.

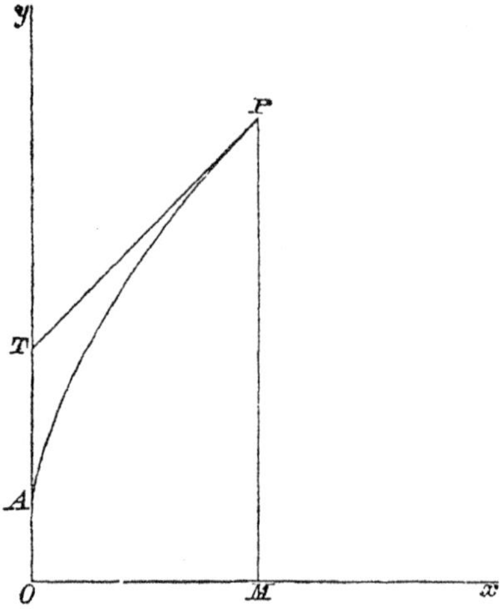

We shall now estimate the resistances; we omit, as before, a certain constant factor.

We imagine a series of zigzags drawn like that of the first particular case of Art. 200.

The resistance on the surface corresponding to AT and TP

$$= a^2 - y_0^2 + \frac{1}{2}(k^2 - a^2).$$

The resistance on the surface which would correspond to the vertical parts of the zigzags along AP

$$= 2\int y\,(dy - dx) = 2\int_{y_0}^{k} y\,dy - 2\int_{0}^{h} y\,dx$$

$$= k^2 - y_0^2 - \text{twice the area } OAPM.$$

The resistance on the surface which would correspond to the inclined parts of the zigzags along AP

$$= 2\int_0^h y\,\frac{1}{2}\,dx = \text{the area } OAPM.$$

Thus the resistance on the surface which would correspond to the whole of the zigzags

$$= k^2 - y_0^2 - \text{the area } OAPM.$$

This exceeds the resistance on the surface corresponding to AT and TP by

$$\frac{1}{2}(k^2 - a^2) - \text{the area } OAPM,$$

that is, by the area ATP; for

$$\frac{1}{2}(k+a)(k-a) \text{ is } \frac{1}{2}(OT + PM)\,OM,$$

and is therefore equal to the area $OTPM$.

By the first particular case of Art. 200, the resistance on the surface generated by the revolution of the curve AP exceeds the resistance on the surface which would be generated by the revolution of the zigzags by $\int \frac{y\,dx\,(dy - dx)^2}{ds^2}$, where the integral extends throughout the curve.

Thus finally the resistance on the surface generated by the revolution of the curve AP exceeds the resistance on the surface generated by the revolution of AT and TP by

$$\int \frac{y\,dx\,(dy - dx)^2}{ds^2} + \text{the area } ATP.$$

202. Next suppose the curve and the tangent produced beyond P; let BS be inclined to the axis of x at an angle of 90°: we shall compare the resistance on the surface generated by the revolution of PS with the resistance on the surface generated by the revolution of PB and BS.

Let x_1, y_1 be the co-ordinates of B; let $SC=b$.

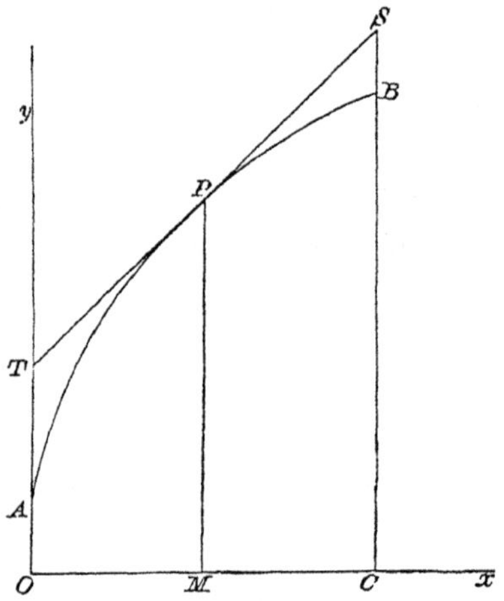

We imagine a series of zigzags drawn like that of the second particular case of Art. 200.

The resistance on the surface corresponding to PS diminished by the resistance on the surface corresponding to BS

$$=\frac{1}{2}(b^2-k^2)-(b^2-y_1^2).$$

The resistance on the surface which would correspond to the vertical parts of the zigzags along PB

$$=2\int y\,(dx-dy)=\text{twice the area } MPBC-(y_1^2-k^2).$$

The resistance on the surface which would correspond to the inclined parts of the zigzags along PB

$$=2\int y\frac{1}{2}\,dx=\text{the area } MPBC.$$

Therefore the resistance on the surface which would correspond to the inclined parts of these zigzags diminished by the resistance on the surface which would correspond to the vertical parts

$$= y_1^2 - k^2 - \text{the area } MPBC.$$

This exceeds the resistance which corresponds to the difference of PS and SB by

$$\frac{1}{2}(b^2 - k^2) - \text{the area } MPBC,$$

that is, by the area of PBS.

By the second particular case of Art. 200 the resistance on the surface generated by the revolution of the curve PB exceeds the resistance on the surface which would correspond to the difference of the vertical and inclined parts of the zigzags by

$$\int \frac{y\,dx\,(dy - dx)^2}{ds^2}.$$

Thus finally the resistance on the surface generated by the revolution of the curve BP exceeds the resistance which corresponds to the difference of PS and SB by

$$\int \frac{y\,dx\,(dy - dx)^2}{ds^2} + \text{the area } PSB.$$

Or, which is the same thing, the resistance on the surface generated by the revolution of the curve PB, together with the resistance on the surface generated by the revolution of BS, exceed the resistance on the surface generated by the revolution of PS by

$$\int \frac{y\,dx\,(dy - dx)^2}{ds^2} + \text{the area } PSB.$$

The investigations of this and the preceding Article are omitted in the paper of the *Quarterly Journal of Mathematics*, to which I have already referred, although they are necessarily required there.

203. The proposition given by R. L. Ellis, to which I referred in Art. 195, is the following: Take the diagram of Art. 202; let AB be an arc of a curve; let TS be a tangent at P, inclined at an angle of 45° to the axis of x; let TA and SB be inclined at an angle of 90° to the axis of x. Then if the figure revolve round the axis of x, the resistance on the surface generated by ST and TA is less than the resistance on the surface generated by SB and BPA.

The demonstration is contained in Arts. 201 and 202.

Mr Ellis established his theorem geometrically. An analytical investigation is given in the *Quarterly Journal of Mathematics* which is unsatisfactory; for it confounds the resistance on the surface corresponding to AT and TP with the resistance on the surface which would correspond to the zigzags along AP. But these two amounts of resistance are not equal; the latter is the greater, as we have seen in Art. 201.

204. The proposition given by R. L. Ellis may also be established in another manner, which indeed resembles his own.

We will confine ourselves to the part PA of the curve, as the remarks made can be easily applied with suitable modifications to the part PB.

The curve may be supposed to be generated by the perpetual intersections of straight lines. Suppose QE and RF to be two consecutive straight lines. Then I shall shew that the resistance on the surface corresponding to FE and EQ is less than the resistance on the surface corresponding to FR and RQ; or, which is the same thing, the resistance on the surface corresponding to FE and ER is less than the resistance on the surface corresponding to FR. It is obvious that if this be true we may pass by a series of changes *at every one of which the resistance is diminished*, from the resistance on the surface corresponding to the AP of Art. 202 to the resistance on the surface corresponding to the AT and TP.

Now the fact that the resistance on the surface corresponding to FE and ER is less than the resistance on the surface corresponding to FR is obvious by the last line of Art. 199. It may be established by a formal use of the method of variations thus:

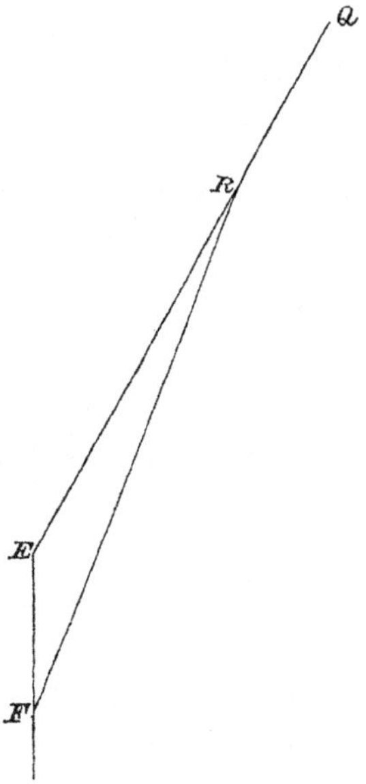

Let y refer to points on FR, and let δy denote the variation by which we pass from FR to ER.

Let $u = \int \frac{yp^3 dx}{1+p^2}$; so that $2\pi u$ measures the resistance on the surface corresponding to FR. The change in passing from FR to FE and ER is therefore

$$2\pi y_0 \delta y_0 + 2\pi \delta u;$$

and
$$\delta u = \int \left\{ \frac{p^3 \delta y}{1+p^2} + \frac{y(3p^2+p^4)\,\delta p}{(1+p^2)^2} \right\} dx$$

$$= \frac{y(3p^2+p^4)\,\delta y}{(1+p^2)^2} + \int \left\{ \frac{p^3}{1+p^2} - \frac{d}{dx} \frac{yp(3p^2+p^4)}{(1+p^2)^2} \right\} \delta y dx.$$

Since p is constant we see that δu reduces to

$$-\left\{ \frac{y(3p^2+p^4)}{(1+p^2)^2} \delta y \right\}_0 - \int \frac{(p^6-p^5+3p^4-p^3)\,\delta y\, dx}{(1+p^2)^2}.$$

Hence $$2\pi y_0 \delta y_0 + 2\pi \delta u$$

$$= 2\pi y_0 \delta y_0 \left\{1 - \frac{(3p^2 + p^4)}{(1 + p^2)^2}\right\}_0 - 2\pi \int \frac{(p^6 - p^5 + 3p^4 - p^3)\, \delta y dx}{(1 + p^2)^2}$$

$$= 2\pi y_0 \delta y_0 \frac{1 - p_0^2}{(1 + p_0^2)^2} - 2\pi \int \frac{(p^6 - p^5 + 3p^4 - p^3)\, \delta y dx}{(1 + p^2)^2}.$$

This is certainly negative if p_0 is not less than unity.

205. We may briefly advert to the problem of finding a solid of *maximum* resistance. In this case we should have the same fundamental equation as in Art. 186. The solid of *greatest* resistance will correspond to the solution given by combining $p = 0$ with $p = \infty$. The generating curve may be supposed to consist of a straight line through A parallel to the axis of x, and the part of the ordinate of B cut off by this straight line.

The same amount of resistance will of course be obtained by a line composed of a series of steps.

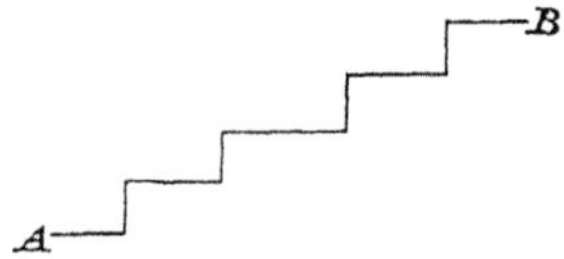

Perhaps besides this solution which gives the greatest resistance a solution might be found to give a *maximum* resistance for admissible variations. Jacobi's theory shews that the value of p^2 in the curve corresponding to (1) of Art. 186 must in that case never be less than 3.

If we require the generating curve to be of given length, there will certainly be some solution for which the resistance is greatest. The equation which will be obtained in the next Article of course applies here.

206. Suppose it were required to determine the solid of revolution of least resistance on the supposition that the generating curve is terminated at two fixed points, and *has a given length*.

By the usual theory we have to seek a minimum of

$$\int \left\{\frac{yp^3}{1 + p^2} + c\sqrt{1 + p^2}\right\} dx,$$

where c is a constant.

This leads to

$$\frac{-2yp^3}{(1+p^2)^2} + \frac{c}{\sqrt{1+p^2}} = \text{a constant};$$

then x may be expressed in terms of p by the relation

$$x = \int \frac{dy}{p}.$$

But the expressions are too complicated to be of any service.

We may however be sure that there must now be some solid of *least* resistance under the condition of given length of the generating curve. Suppose the given length happened to be exactly the length of the curve in Art. 190; then that curve would give *a* minimum solution of the present problem. If we impose also the condition that p is to be always of one sign, then in the case in which the solution of Art. 190 is the only solution of the problem without the condition of given length, it will be the only solution of the present problem if the given length happens to coincide with the length thus obtained. Hence we have a discontinuous solution of the present problem in certain cases.

207. We proceed to another variety of the problem of a solid of minimum resistance.

A solid of revolution is to be formed on a given base, so as to have a given surface and to experience a minimum resistance when it moves through a fluid in the direction of its axis.

Take the axis of x for that of revolution, and make y the independent variable. We have then to find a minimum of

$$\int_0^b \left\{\frac{y}{1+\varpi^2} + cy\sqrt{(1+\varpi^2)}\right\} dy,$$

where ϖ stands for $\frac{dx}{dy}$ and c is a constant.

Denote the integral by u; then

$$\delta u = \int_0^b y \left\{\frac{c\varpi}{\sqrt{(1+\varpi^2)}} - \frac{2\varpi}{(1+\varpi^2)^2}\right\} \delta\varpi\, dy$$

$$+ \int_0^b y \left\{\frac{c}{2(1+\varpi^2)^{\frac{3}{2}}} + \frac{3\varpi^2-1}{(1+\varpi^2)^3}\right\} (\delta\varpi)^2\, dy.$$

The term of the first order is transformed in the usual way, and is made to vanish by the supposition

$$y\varpi \left\{\frac{c}{\sqrt{(1+\varpi^2)}} - \frac{2}{(1+\varpi^2)^2}\right\} = \text{a constant};$$

and this constant must be zero since the generating curve is supposed to meet the axis of x. We have then to form a solution consisting of a straight line or straight lines, which are furnished by

$$y = 0, \quad \text{or } \varpi = 0, \quad \text{or } \frac{c}{\sqrt{(1+\varpi^2)}} = \frac{2}{(1+\varpi^2)^2}.$$

The last of these gives $\varpi^2 = \left(\frac{2}{c}\right)^{\frac{2}{3}} - 1.$

Thus if $BC = b$, and we draw BA so that the surface generated by the revolution of BA may have the given value, this conical

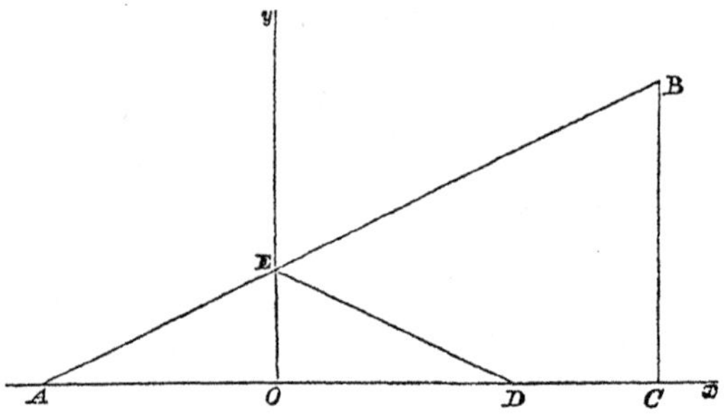

surface satisfies the conditions for a minimum. For the term of the first order in δu vanishes, and the term of the second order becomes $\int_0^b \frac{3y\varpi^2}{(1+\varpi^2)^3}(\delta\varpi)^2\,dy$, which is positive.

208. If ED is equal to EA we may take the discontinuous locus formed of BE and ED instead of that formed by BA. Thus if we propose that the generating curve shall pass through a fixed point D on the axis we may suppose that our generating curve is composed of BA and AD, or of BE and ED: the surface and the resistance are the same in the two cases.

The discontinuity which exists when we take BE and ED for the generating line is like that of Art. 9, and arises in the same way: the equation here from which ϖ is found gives us two values numerically equal but of opposite signs.

It is obvious that we may increase the number of zigzags without changing either the area of the surface or the amount of the resistance; and so the generating curve can always if we please be comprised between the ordinate at B and any other fixed ordinate.

209. We observe that the case $\varpi = 0$ presents itself among those to be examined in Art. 208. This gives great variety to the figure which has the property of a minimum.

For example, suppose DE parallel to the axis of y, and FE and DB parallel straight lines. Let the lengths of the lines be such

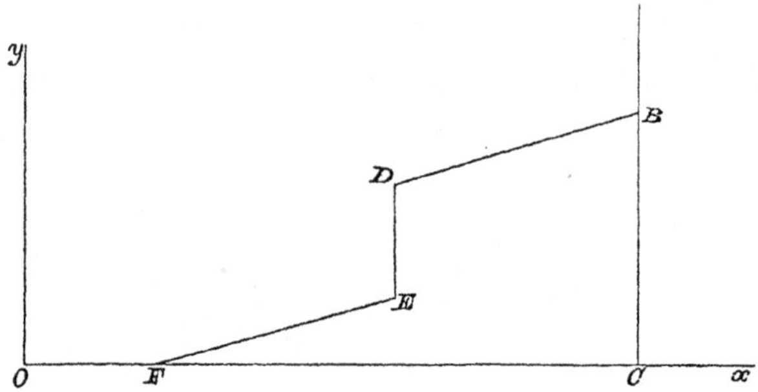

that if $FEDB$ revolve round the axis of x the surface generated has the given value. Then the figure thus formed has the minimum property; that is, any figure obtained from this by an admissible variation, and generating an equal surface, will correspond to a greater resistance, provided the variation is not limited to the part DE alone.

For the term of the first order in δu vanishes, and the term of the second order vanishes for the part DE, and for the rest of the figure retains the form given to it in Art. 207; and this is positive whatever may be the value of ϖ.

210. But such a figure as that in the preceding Article will not correspond to the *least* resistance. To shew this we have only to consider the following proposition:

A surface is generated by the revolution of a straight line PS round the axis of x; another surface is generated by the revolution

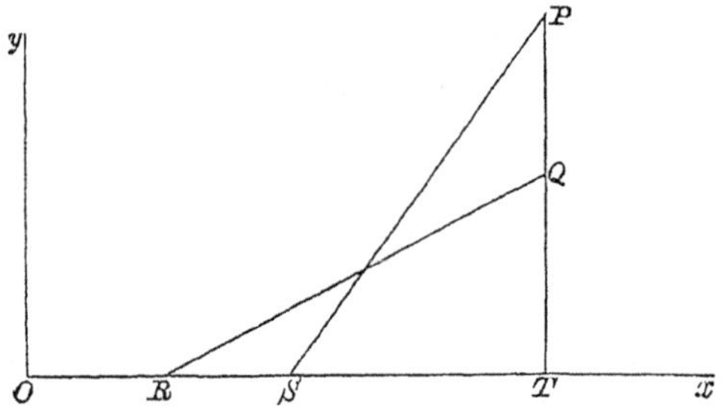

of the composite line PQ and QR; supposing the areas of the two surfaces equal, the resistance on the latter is the greater.

Let $PT = b$, $QT = r$, $PST = \alpha$, $QRT = \beta$.

Since the surfaces are equal we have

$$\frac{b^2}{\sin\alpha} = b^2 - r^2 + \frac{r^2}{\sin\beta}; \quad \text{therefore } r^2 = \frac{b^2(1-\sin\alpha)\sin\beta}{(1-\sin\beta)\sin\alpha}.$$

The resistance corresponding to PS varies as $b^2\sin^2\alpha$; the resistance corresponding to PQR varies as $r^2\sin^2\beta + b^2 - r^2$: the latter $= b^2\sin^2\alpha + b^2(1-\sin^2\alpha) - r^2(1-\sin^2\beta)$

$$= b^2\sin^2\alpha + b^2\left\{1 - \sin^2\alpha - \frac{\sin\beta(1-\sin\alpha)(1+\sin\beta)}{\sin\alpha}\right\}$$

$$= b^2\sin^2\alpha + \frac{b^2(1-\sin\alpha)}{\sin\alpha}\left\{(1+\sin\alpha)\sin\alpha - (1+\sin\beta)\sin\beta\right\}$$

$$= b^2\sin^2\alpha + \frac{b^2(1-\sin\alpha)}{\sin\alpha}(\sin\alpha - \sin\beta)(1 + \sin\alpha + \sin\beta).$$

It is obvious that this is greater than $b^2\sin^2\alpha$.

211. Thus it is clear that we cannot obtain the solid of *least* resistance if we make any use of the solution $\varpi = 0$.

As to the solution $y = 0$ this would correspond to $\varpi = \infty$, and the investigation which has been given is not very satisfactory in this case. But on changing the independent variable from y to x

we shall see that $y=0$ corresponds to a minimum. But this result is of no consequence, as the surface and the resistance both become zero when $y=0$; so that it makes no difference what portion of the axis of x is comprised in our solution.

Hence it follows that to obtain the solid of *least* resistance we must take the solution of Art. 207.

212. Suppose in Art. 210 that Q is very near to P; still the conclusion holds that the resistance on PQ and QR is greater than that on PS. This may at first appear inconsistent with the statement of Art. 209, that the diagram there given corresponds to a *minimum;* for it is clear that by such a change as consists in taking PS instead of PQ and QR the resistance is diminished. But it must be observed that such a change cannot be made by an *admissible* variation; in passing from PQ to PS although the variation in x may be infinitesimal, that in ϖ will not be infinitesimal.

213. Suppose in this problem that we require the generating curve to terminate at fixed points, and also impose the condition that $\frac{dy}{dx}$ shall never change sign; the following is the solution:

Let A and B denote the fixed points, A being on the axis of revolution. Draw BC and BF perpendicular to the axes.

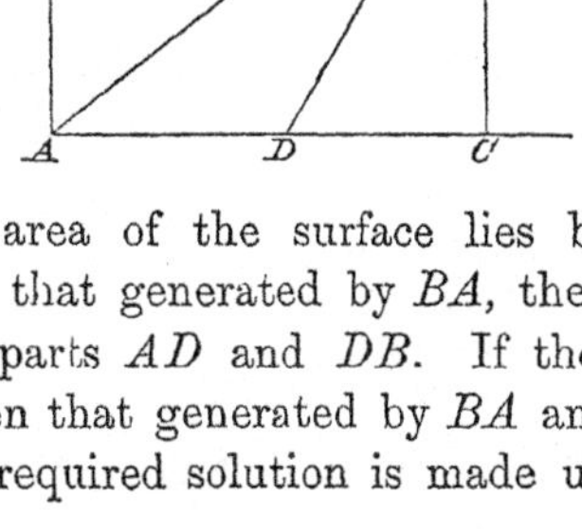

If the given area of the surface lies between that generated by BC and that generated by BA, the required solution is made up of two parts AD and DB. If the given area of the surface lies between that generated by BA and that generated by AF and FB, the required solution is made up of two parts AE

and EB. In the former case we in fact combine $y=0$ and $\frac{c}{\sqrt{(1+\varpi^2)}}=\frac{2}{(1+\varpi^2)^2}$; in the latter case we combine $\varpi=0$ and $\frac{c}{\sqrt{(1+\varpi^2)}}=\frac{2}{(1+\varpi^2)^2}$.

We may observe that it can be immediately shewn that the area of the surface in the second case increases as AE increases. For take the general expression for δS given in Art. 84, and put $q=0$; thus we obtain for the variation produced by a change of a straight line BE to a slightly higher position

$$-2\pi\left\{\frac{py\,\delta y}{\sqrt{(1+p^2)}}\right\}_0+2\pi\int\frac{\delta y\,dx}{\sqrt{(1+p^2)}};$$

the latter term is positive, the former term is less than $2\pi\,(y\delta y)_0$, which is the increment of the surface generated by AE.

214. Let us now briefly advert to the problem in which the resistance is required to be a *maximum*, the area of the surface being given as before.

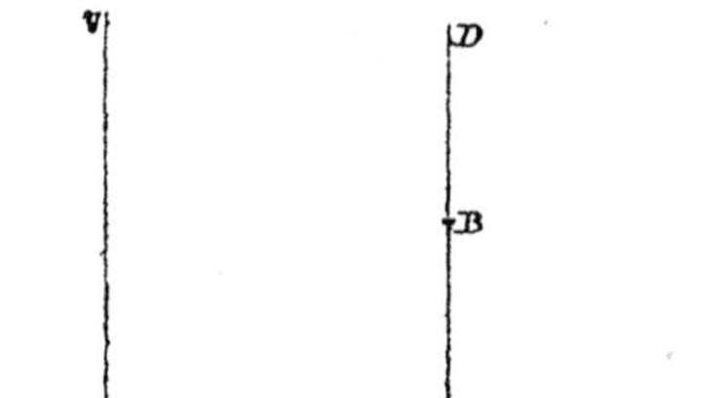

It is evident that the greatest resistance is obtained by supposing the surface generated by a straight line DBC at right angles to the axis of such a length that the area of the circle thus obtained may be equal to the given area.

This result may be extracted too from the formulæ; the solution must be considered to be $\varpi=0$. Let CD be denoted by y_1; then we must remember that y_1 is variable, and so to the value of δu in Art. 207 we must now add

$$\left\{\frac{y}{1+\varpi^2} + cy\sqrt{(1+\varpi^2)}\right\}_1 \delta y_1,$$

that is $$y_1(1+c)\,\delta y_1 \text{ since } \varpi = 0.$$

To make this vanish we should require $c = -1$; and then the term in δu of the second order is negative.

If however the generating line is not allowed to have any ordinate greater than that of B we must take a different solution.

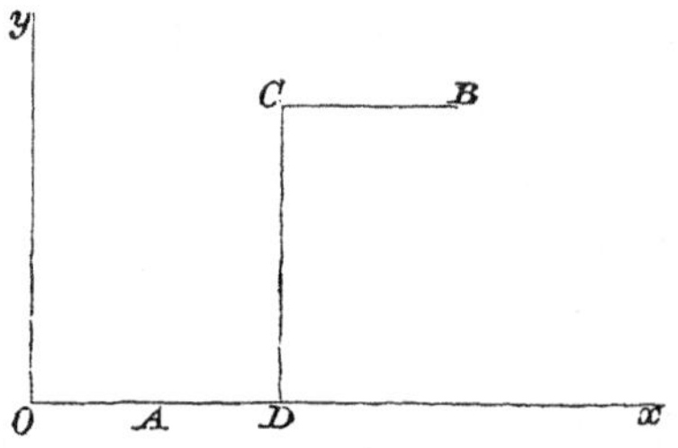

Draw BC parallel to the axis of x and CD perpendicular to it; let BC be of such length that the surface generated by the revolution of BC and CD may have the given area: then the solution may be considered to be made up of BC and CD.

The part CD corresponds to $\varpi = 0$; the part CB corresponds to $\varpi = \infty$, which with $c = 0$ may be considered to satisfy the fundamental equation of Art. 207. It would be unsafe to rely upon the investigation on account of the infinite value of ϖ; but it is obvious that this figure gives us the greatest possible amount of resistance.

CHAPTER X.

SOLID OF MINIMUM RESISTANCE WITH GIVEN VOLUME.

215. THE following interesting variety of the problem of the solid of minimum resistance has been recently considered: a solid of revolution is to be formed on a given base with a given volume so as to experience a minimum resistance when it moves through a fluid in the direction of its axis. See *Philosophical Magazine* for November, 1867.

I borrow little more than the enunciation of the problem; in fact the discussion which will now be given is almost entirely new: all that relates to the interpretation of the results, and to discontinuity, is of course here given for the first time.

216. Let Ox denote the axis of revolution, AB the generating curve, BC the radius of the given base.

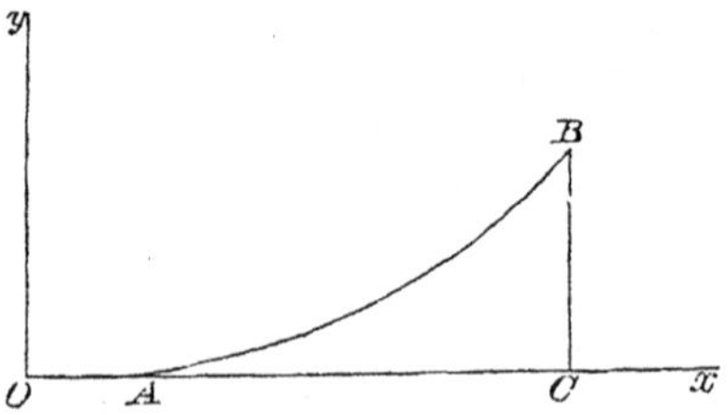

Take the origin O at any point of the axis; let $OC = a$, and $BC = b$; these may be considered known quantities.

Let $OA = x_0$ which is not known.

By the usual theory we seek for a minimum of

$$\int_{x_0}^{a} \left(\frac{yp^3}{1+p^2} + 2\lambda y^2\right) dx,$$

where λ is a constant at present undetermined.

By putting the variation of this to the first order equal to zero, we obtain in the usual way

$$\frac{yp^3}{1+p^2} + 2\lambda y^2 = \frac{yp\,(3p^2+p^4)}{(1+p^2)^2} + \text{ a constant.}$$

The constant here must be zero since the curve meets the axis of x. Hence

$$\lambda y = \frac{p^3}{(1+p^2)^2};$$

or putting $\frac{1}{c}$ for λ, we have

$$y = \frac{cp^3}{(1+p^2)^2} \quad \text{..........................} \; (1).$$

Thus (1) determines the generating curve; we shall shew that this curve is a hypocycloid.

Put $p = \tan\phi$; then

$$y = c \sin^3\phi \cos\phi;$$

therefore $$\frac{dy}{dx} = c\,(3\sin^2\phi\cos^2\phi - \sin^4\phi)\,\frac{d\phi}{dx};$$

therefore $$\frac{dy}{d\phi} = c\,(3\cos^2\phi - \sin^2\phi)\sin^2\phi,$$

and $$\frac{dx}{d\phi} = c\,(3\cos^2\phi - \sin^2\phi)\sin\phi\cos\phi.$$

Hence, squaring and adding, we have

$$\frac{ds}{d\phi} = c\,(3\cos^2\phi - \sin^2\phi)\sin\phi = c\sin 3\phi,$$

where s denotes the arc of the curve; therefore

$$s = \text{constant} - \frac{c}{3}\cos 3\phi.$$

Hence the curve is a hypocycloid in which the radius of the moving circle is one-third of the radius of the fixed circle: see Todhunter's *Integral Calculus*, Art. 112.

When $y=0$, we have either $\phi=0$ or $\phi=\frac{\pi}{2}$: at present we shall discuss the former case.

217. We suppose then that $\phi=0$ when $y=0$, and we measure s from that point; thus

$$s=\frac{c}{3}(1-\cos 3\phi).$$

Let ϕ_1 be the value of ϕ when $y=b$; then to determine c and ϕ_1 we have the following equations:

$$b=c\sin^3\phi_1\cos\phi_1,$$

$$\pi\int_{x_0}^{a} y^2\,dx = \text{the given volume};$$

the latter equation may be written

$$\pi c^3\int_0^{\phi_1}\sin^6\phi\sin 3\phi\cos^3\phi\,d\phi = \text{the given volume}.$$

Effecting the integration and substituting for c, we find that the expression on the left-hand side is

$$\frac{\pi b^3\left(\frac{3}{8}-\frac{7}{10}\sin^2\phi_1+\frac{1}{3}\sin^4\phi_1\right)}{\sin\phi_1\cos^3\phi_1}.$$

It is obvious that the value of this expression can be made as great as we please by taking ϕ_1 small enough: but the value cannot be made as small as we please, for of course it is greater than

$$\pi b^3\left(\frac{3}{8}-\frac{7}{10}\sin^2\phi_1+\frac{1}{3}\sin^4\phi_1\right);$$

and it may be shewn that this is always positive, and has its least value when $\phi_1=\frac{\pi}{2}$, namely $\frac{\pi b^3}{120}$.

Hence the solution we are considering becomes inapplicable when the given volume is less than a certain definite limit which may be easily assigned. For put the expression found above for the volume in terms of $\tan\phi_1$; it will be found that the expression becomes

$$\pi b^3\left(\frac{\tan^3\phi_1}{120}+\frac{\tan\phi_1}{20}+\frac{3}{8\tan\phi_1}\right);$$

by differentiating with respect to ϕ_1 we find that this expression constantly diminishes as ϕ_1 increases from 0 up to $\frac{\pi}{3}$; the least value is when $\phi_1=\frac{\pi}{3}$; the value is then

$$\frac{\pi b^3\sqrt{3}}{5}.$$

218. Let us now consider the term of the second order in the variation of u where u denotes the integral in Art. 216.

Put $\phi(p)$ for $\frac{p^3}{1+p^2}$.

We observe that since the limit x_0 is not fixed, we must attribute to it a change or variation dx_0; in consequence of this there occurs in δu the term

$$\{y\,\phi'(p)\,\delta y\}_0-\int_{x_0}^{x_0+dx_0}\{y\,\phi(p)+2\,\lambda y^2\}\,dx,$$

which we have not yet regarded. This term is of the first order in *appearance;* but as y, $\phi(p)$, and $\phi'(p)$ all vanish when $x=x_0$ we may consider that it is not even of the second order in value. There is a relation between δy_0 and dx_0; see Todhunter's *Integral Calculus,* Art. 359: but it is here of no importance.

The term of the second order in the integral is

$$\int_{x_0}^{a}\left\{\phi'(p)\,\delta y\,\delta p+\frac{y}{2}\,\phi''(p)\,(\delta p)^2+2\lambda\,(\delta y)^2\right\}dx,$$

and by transforming this as in Art. 26 we obtain

$$\frac{1}{2}\int_{x_0}^{a} [\phi''(p)\{y(\delta p)^2 - y''(\delta y)^2\} + 4\lambda(\delta y)^2]\, dx;$$

for the term outside the integral sign involving $(\delta y_0)^2$ vanishes.

But from (1)

$$\lambda y = \frac{p^3}{(1+p^2)^2};$$

hence, by differentiation,

$$\lambda p = \frac{3p^2 - p^4}{(1+p^2)^3} y'',$$

that is
$$2\lambda = y''\phi''(p);$$

and so the term of the second order in δu becomes

$$\frac{1}{2}\int_{x_0}^{a} [\phi''(p)\, y\, (\delta p)^2 + 2\lambda(\delta y)^2]\, dx:$$

or we may put it in the form

$$\frac{1}{2}\int_{x_0}^{a} \phi''(p)\{y(\delta p)^2 + y''(\delta y)^2\}\, dx.$$

Now $\phi''(p)$ is positive as long as p^2 is less than 3; and thus we see that if ϕ_1 does not exceed $\frac{\pi}{3}$, the term of the second order in δu is essentially positive: if we take the second form we must remember that y'' is positive, as the arc which we have to consider is convex to the axis of x.

219. Thus if the given volume is not less than $\frac{\pi b^3\sqrt{3}}{5}$, we have obtained a solid of *minimum* resistance. We must not say

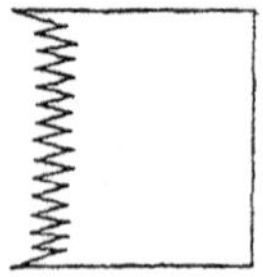

that we have obtained the solid of *least* resistance, for by using a zigzag boundary we could make the resistance as small as we please for any given volume. All that we can assert is that the resistance is less for the solid which we have obtained than it would be for any solid which could be obtained from this by infinitesimal variations of y and p.

220. If however we impose the condition that p shall have always the same sign we may take it as obvious that there must be some solid of *least* resistance; this solid then must be either that which we have obtained or must be furnished by some discontinuous solution: it is certain that a discontinuous solution must exist in some cases, for we have not as yet even a minimum solution in the case in which the given volume is less than the limit which we have specified.

221. We proceed then to seek for this discontinuous solution.

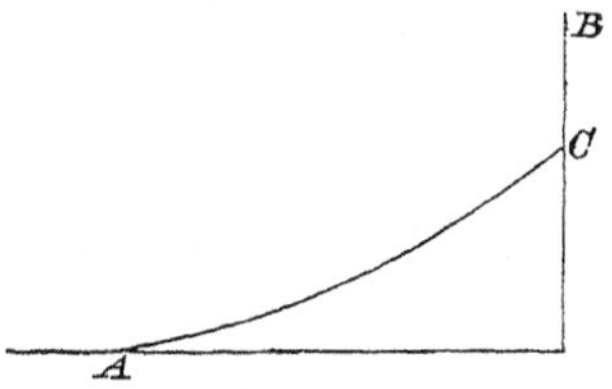

The only supposition which suggests itself is that the solution consists of a curve AC, and CB a portion of the ordinate at B. As $p = \infty$ does not present itself as a solution of the fundamental equation of Art. 216, we do not seem to be very naturally led to this supposition. We shall at present however verify that we do thus obtain a solution for some cases; and in a subsequent Article we shall remove the apparently arbitrary character of the supposition.

222. Let y_1 be the ordinate of C. The resistance on AC is measured by $2\pi \int_{x_0}^{a} \frac{yp^3 dx}{1+p^2}$; and that on BC by $\pi (b^2 - y_1^2)$. The volume is $\pi \int_{x_0}^{a} y^2 dx$. Hence by the usual theory we require a minimum of

$$b^2 - y_1^2 + 2\int_{x_0}^{a}\left(\frac{yp^3}{1+p^2} + 2\lambda y^2\right) dx.$$

Call this u; then form δu to the first order.

By making the part under the integral sign vanish we obtain as before equation (1) for the curve AC. Outside the integral sign we have

$$-2y_1\delta y_1 + 2\left\{\frac{3p^2+p^4}{(1+p^2)^2}\right\}_1 y_1\delta y_1;$$

to make this vanish we must have $p_1 = 1$, so that the curve meets CB at an angle of 45°.

The term of the second order which arises from the variation of $-y_1^2$ is balanced by an equal term of contrary sign arising from the integral in u: see Art. 190. Thus we are left with a term of the second order just double of that which we had at the end of Art. 218, and therefore essentially positive. Thus we have a minimum, provided satisfactory values can be found for the constants which we shall now consider.

223. To determine the constants we have

$$y_1 = \frac{c}{4},$$

$$\pi c^3 \int_0^{\frac{\pi}{4}} \sin^6\phi \sin 3\phi \cos^3\phi d\phi = \text{the given volume.}$$

The latter becomes by effecting the integration

$$\frac{13\pi c^3}{1920} = \text{the given volume.}$$

Moreover we must have y_1 not greater than b, that is $\frac{c}{4}$ not greater than b. Hence the *greatest* admissible value of the volume is when $c = 4b$, and is therefore $\frac{13\pi b^3}{30}$.

224. Thus the following results hold for the solid of *least* resistance on a given base and with a given volume, with the condition that p is to be always of the same sign.

I. If the given volume is less than $\frac{\pi b^3 \sqrt{3}}{5}$ we must take the discontinuous solution.

II. If the given volume is greater than $\frac{13\pi b^3}{30}$ we must take the continuous solution.

III. If the given volume lies between $\frac{\pi b^3 \sqrt{3}}{5}$ and $\frac{13\pi b^3}{30}$ we must determine which of the two solutions gives the least resistance and take that solution. We shall presently shew that the discontinuous solution is that which gives the least resistance in this case.

225. It is easy to give expressions for the amount of the resistance.

$$2\pi \int \frac{p^3 y\, dx}{1+p^2} = 2\pi \int \frac{p^2 y\, dy}{1+p^2} = 2\pi c \int \frac{p^5 dy}{(1+p^2)^3}.$$

By Art. 216 this may be transformed to

$$2\pi c^2 \int \cos\phi \sin^7\phi\, (3 - 4\sin^2\phi)\, d\phi\,;$$

integrating between the limits 0 and ϕ_1 we obtain

$$2\pi c^2 \left(\frac{3}{8}\sin^8\phi_1 - \frac{4}{10}\sin^{10}\phi_1\right).$$

Then for the continuous solution we substitute for c from the equation

$$b = c\sin^3\phi_1 \cos\phi_1\,;$$

thus the resistance

$$= \frac{\pi b^2}{\cos^2\phi_1}\left(\frac{3}{4}\sin^2\phi_1 - \frac{4}{5}\sin^4\phi_1\right);$$

and ϕ_1 is to be found from the equation

$$\pi b^3 f(\phi_1) = \text{the given volume},$$

where $f(\phi_1)$ is put for

$$\frac{\frac{3}{8} - \frac{7}{10}\sin^2\phi_1 + \frac{1}{3}\sin^4\phi_1}{\sin\phi_1 \cos^3\phi_1}$$

that is, for

$$\frac{1}{120}\tan^3\phi_1 + \frac{1}{20}\tan\phi_1 + \frac{3}{8}\cot\phi_1.$$

For the discontinuous solution we put $\phi_1 = \frac{\pi}{4}$, and to obtain the whole resistance we add $\pi(b^2 - y_1^2)$; thus the whole resistance

$$= \frac{7\pi c^2}{320} + \pi\left(b^2 - \frac{c^2}{16}\right) = \pi b^2 - \frac{13\pi c^2}{320},$$

where c is to be found from the equation

$$\frac{13\pi c^3}{1920} = \text{the given volume.}$$

226. Let us take for an example the case in which the given volume is $\frac{\pi b^3\sqrt{3}}{5}$.

In this case in the continuous solution we put $\phi_1 = \frac{\pi}{3}$; and we find the resistance to be $\frac{9\pi b^2}{20}$.

In the discontinuous solution we put

$$\frac{13\pi c^3}{1920} = \frac{\pi b^3\sqrt{3}}{5};$$

this gives

$$c^2 = b^2 \times \frac{48}{13}\sqrt[3]{52};$$

and thus the resistance is

$$\pi b^2\left(1 - \frac{3}{20}\sqrt[3]{52}\right);$$

we find this to be approximately $\pi b^2 \times \cdot 44012$.

Therefore the resistance is less for the discontinuous solution than for the continuous solution.

When the given volume is $\frac{13\pi}{30}b^3$ the so-called discontinuous solution coincides with the continuous: we have $\phi_1 = \frac{\pi}{4}$, and the resistance is $\frac{7\pi b^2}{20}$.

227. We shall now shew that both for the continuous and the discontinuous solution the resistance decreases as the given volume increases; and that when both solutions are applicable the resistance is less for the discontinuous solution than for the other.

Let V denote the volume; let R denote the resistance for the continuous solution and S for the discontinuous solution corresponding to the same volume V.

Then $$R = \frac{\pi b^2 \left\{\frac{3}{4}(1-\cos^2\phi_1) - \frac{4}{5}(1-\cos^2\phi_1)^2\right\}}{\cos^2\phi_1}$$

$$= \frac{\pi b^2}{\cos^2\phi_1}\left\{-\frac{1}{20} + \frac{17}{20}\cos^2\phi_1 - \frac{4}{5}\cos^4\phi_1\right\},$$

hence we obtain

$$\frac{dR}{d\phi_1} = \frac{\pi b^2 \sin\phi_1}{10\cos^3\phi_1}(16\cos^4\phi_1 - 1).$$

Also $$V = \pi b^3\left\{\frac{1}{120}\tan^3\phi_1 + \frac{1}{20}\tan\phi_1 + \frac{3}{8}\cot\phi_1\right\};$$

hence we obtain

$$\frac{dV}{d\phi_1} = \frac{\pi b^3}{40\sin^2\phi_1\cos^4\phi_1}(1 - 16\cos^4\phi_1).$$

Thus $$\frac{dR}{dV} = -\frac{4}{b}\sin^3\phi_1\cos\phi_1$$

$$= -\frac{4}{\gamma} \text{ by Art. 217,}$$

where γ is used instead of c for the parameter of the hypocycloid: we shall reserve c for the parameter of the hypocycloid in the discontinuous solution.

And $$S = \pi b^2 - \frac{13\pi c^2}{320};$$

therefore $$\frac{dS}{dc} = -\frac{26\pi c}{320}.$$

Also $$V=\frac{13\pi c^3}{1920};$$

therefore $$\frac{dV}{dc}=\frac{39\pi c^2}{1920}.$$

Thus $$\frac{dS}{dV}=-\frac{4}{c}.$$

These results shew that both R and S diminish as V increases.

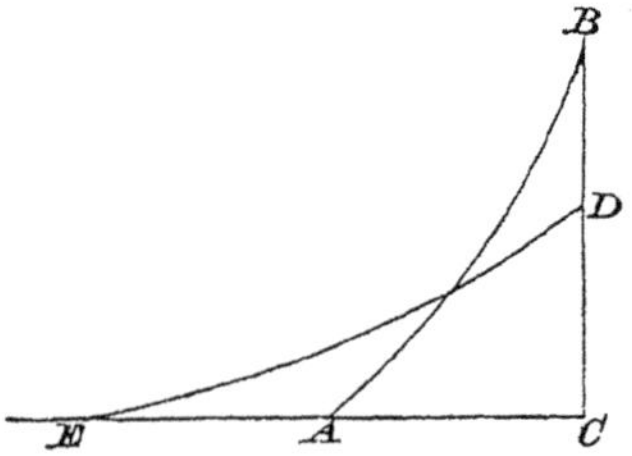

Let BA represent the generating curve for the continuous solution, and DE for the discontinuous solution corresponding to the same volume. These curves then are similar; but it is obvious that DE is on the *larger scale*. Thus c is greater than γ, and so γ is less than c.

Therefore $\frac{dS}{dV}$ is numerically less than $\frac{dR}{dV}$. Now we know that when $V=\frac{\pi b^3\sqrt{3}}{5}$ we have S less than R; and when $V=\frac{13\pi b^3}{30}$ we have $S=R$. It follows that for all intermediate values of V we must have S less than R. For if R could be equal to S for any intermediate value of V, then as V increased R would decrease more rapidly than S; and thus R could not be equal to S when V became equal to $\frac{13\pi b^3}{30}$.

228. It will be seen that we obtained in the preceding Article for $\frac{dR}{dV}$ and $\frac{dS}{dV}$ results of the *same form*; this at first sight

may appear strange: but the reason for it can be easily assigned. Suppose in the preceding diagram we were to draw just above DE the hypocycloid for the discontinuous solution corresponding to a slightly increased volume: let it be FG.

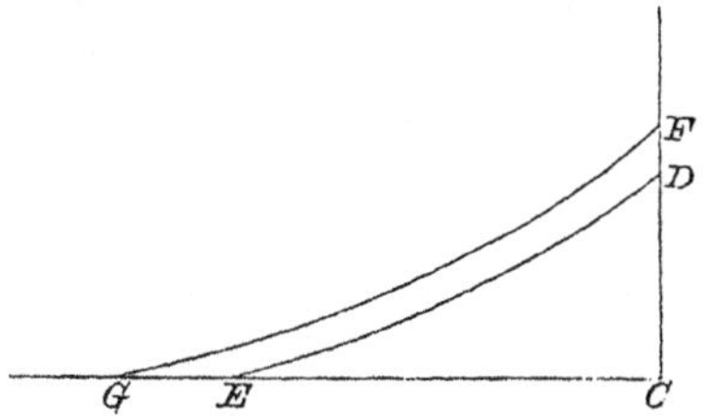

Let S denote the resistance on ED and the corresponding part of CD produced; and let V denote the volume. Then the principle on which the problem in the Calculus of Variations is solved is that of making $\delta(S + 4\lambda V) = 0$: see Art. 222. Thus $\delta S = -4\lambda\delta V$. This is true for all variations, and therefore for the variation by which we pass from DE to FG. This result expressed in other notation becomes $\frac{dS}{dV} = -4\lambda = -\frac{4}{c}$.

Similarly we account for the result which is expressed by $\frac{dR}{dV} = -\frac{4}{\gamma}$.

229. As another numerical example suppose $\tan\phi_1 = \frac{3}{2}$. Then, $f(\phi_1)$ having the meaning of Art. 225, we find that $f(\phi_1) = \frac{113}{320}$. To determine c we have $\frac{13c^3}{1920} = \frac{113}{320}b^3$; so that $c = \left(\frac{6 \times 113}{13}\right)^{\frac{1}{3}} b$.

It will be found that for the continuous solution the resistance is

$$\frac{459\pi b^2}{1040} = \pi b^2 \times \cdot 44134\ldots\ldots;$$

and for the discontinuous solution the resistance is

$$\pi b^2 \left\{1 - \frac{(36 \times 13 \times 113 \times 113)^{\frac{1}{3}}}{320}\right\} = \pi b^2 \times \cdot 43291\ldots\ldots$$

230. Let us now briefly consider the other case which presented itself in Art. 216, namely, that in which we suppose $\phi = \frac{\pi}{2}$ when $y = 0$.

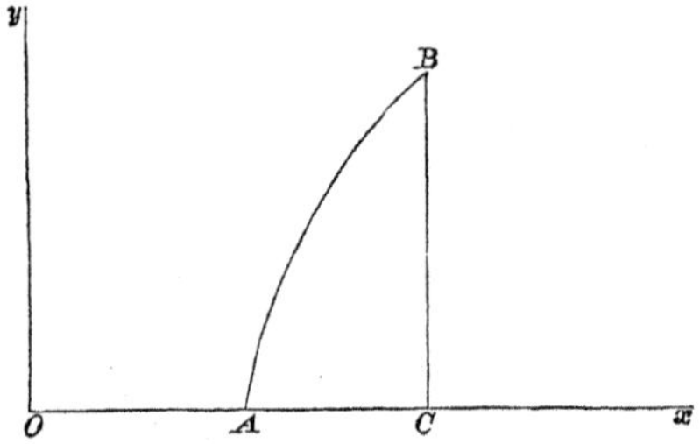

If we measure s from this point we have

$$s = -\frac{c}{3} \cos 3\phi.$$

As before we shall have

$$b = c \sin^3 \phi_1 \cos \phi_1.$$

And ϕ_1 must lie between $\frac{\pi}{3}$ and $\frac{\pi}{2}$ in order that p may be of invariable sign which we have assumed as a condition.

As before we have

$$\frac{\pi b^3 \int_{\frac{\pi}{2}}^{\phi_1} \sin^6 \phi \sin 3\phi \cos^3 \phi \, d\phi}{\sin^9 \phi_1 \cos^3 \phi_1} = \text{the given volume.}$$

To effect the integration conveniently in this case we put the integral in the form

$$\int_{\frac{\pi}{2}}^{\phi_1} \cos^3 \phi \, (1 - \cos^2 \phi)^3 \, (4 \cos^2 \phi - 1) \sin \phi \, d\phi \,;$$

hence taking the integral between the limits we get

$$\frac{\pi b^3 \cos \phi_1}{\sin^9 \phi_1} \left\{ \frac{1}{3} \cos^8 \phi_1 - \frac{13}{10} \cos^6 \phi_1 + \frac{15}{8} \cos^4 \phi_1 - \frac{7}{6} \cos^2 \phi_1 + \frac{1}{4} \right\}$$

$$= \text{the given volume.}$$

It is obvious that the value of the expression on the left-hand side can be made as small as we please by taking ϕ_1 near enough to $\frac{\pi}{2}$; but the value of the expression cannot be made as great as we please, ϕ_1 being restricted to lie between $\frac{\pi}{3}$ and $\frac{\pi}{2}$. In fact, denoting the expression by $\pi b^3 F(\phi_1)$, we shall find that

$$F(\phi) = -\frac{1}{120} \cot^9 \phi - \frac{1}{20} \cot^7 \phi - \frac{1}{8} \cot^5 \phi - \frac{1}{6} \cot^3 \phi + \frac{1}{4} \cot \phi,$$

and that $\dfrac{dF(\phi)}{d\phi} = \dfrac{3 \cot^2 \phi - 1}{40 \sin^2 \phi} (\cot^6 \phi + 5 \cot^4 \phi + 10 \cot^2 \phi + 10)$;

so that $F(\phi)$ decreases continually as ϕ increases from $\frac{\pi}{3}$ to $\frac{\pi}{2}$. Thus the greatest admissible volume corresponds to $\phi_1 = \frac{\pi}{3}$, and is $\dfrac{\pi b^3 \times 217}{1215 \sqrt{3}}$.

231. Since in this case p is infinite when $y = 0$, our investigation of the variation of the integral to be considered is not satisfactory, and it will be better to take y as the independent variable. The investigation will be given presently and will lead to the conclusion that we have now a *maximum*. But of course this result must be understood with due restriction. We must not suppose that we have thus a solid of *greatest* resistance; for the greatest resistance is obviously when the solid is a cylinder, or, which amounts to the same thing, when the solid has the step-shaped boundary indicated in Art. 205. The statement merely means that the resistance is a maximum with respect to any *admissible* variation. A greater resistance can be obtained immediately by replacing the boundary by a figure like that in

Art. 205; this is however not an admissible variation, because although the value of δy might be made everywhere infinitesimal, yet that of δp could not.

232. Let us now give the formulæ for solving the problem when y is taken for the independent variable. Let ϖ stand for $\frac{dx}{dy}$. Then the resistance $= 2\pi \int_0^{y_1} \frac{y\,dy}{1+\varpi^2}$,

and the volume $= \pi \int_0^{y_1} y^2 \varpi\, dy$.

Thus we seek for a maximum or minimum of

$$\int_0^{y_1} \left(\frac{y}{1+\varpi^2} + 2\lambda y^2 \varpi\right) dy,$$

where λ is a constant.

The complete variation to the second order, supposing the upper limit changed from y_1 to $y_1 + dy_1$ is

$$\int_0^{y_1} \left\{\frac{-2y\varpi}{(1+\varpi^2)^2} + 2\lambda y^2\right\} \delta\varpi\, dy + \int_0^{y_1} \frac{(3\varpi^2-1)\, y\, (\delta\varpi)^2}{(1+\varpi^2)^3}\, dy$$

$$+ \int_{y_1}^{y_1+dy_1} \left\{\frac{y}{1+(\varpi+\delta\varpi)^2} + 2\lambda y^2 (\varpi + \delta\varpi)\right\} dy.$$

The first term is made to vanish in the usual way by putting

$$2\lambda y^2 - \frac{2y\varpi}{(1+\varpi^2)^2}$$

equal to a constant, which constant must be zero. This of course leads to the same results as in Art. 216. The case which we proceeded to discuss in Art. 217 is best treated as it was there by taking x as the independent variable. The present investigation will be suitable instead of that which began in Art. 230, as it avoids the infinite quantity which there occurred.

We see, however, that if we do not suppose $dy_1 = 0$, we do not make the term of the first order in the variation vanish; but if we take $dy_1 = 0$ the term of the first order does vanish, and the variation reduces to

$$\int_0^{y_1} \frac{(3\varpi^2-1)\, y\, (\delta\varpi)^2}{(1+\varpi^2)^3}\, dy,$$

which is negative, because ϖ^2 is less than $\frac{1}{3}$ throughout the integration. Hence we have a maximum for *admissible* variations.

233. There is a peculiarity in the investigation of the preceding Article which requires notice; a particular variation is inadmissible which might appear at first sight to be admissible.

Let BA be a curve; let BD be parallel to the axis of x, and suppose it indefinitely small: let DE be another curve.

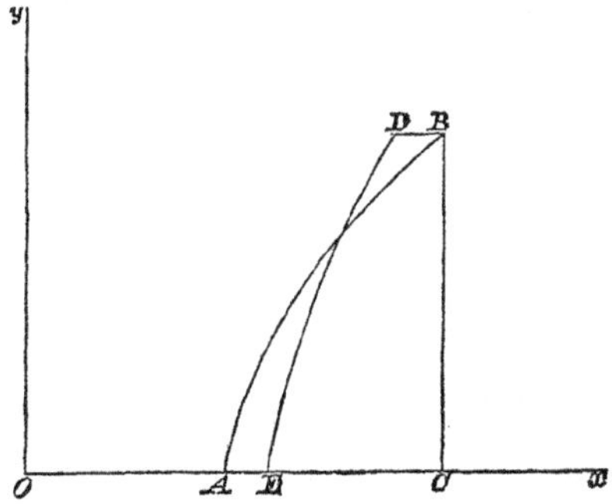

Then we could pass from such a curve as BA to such a boundary as BD and DE by means of infinitesimal changes δx and $\delta\varpi$; though we could not by infinitesimal changes in y and p: but nevertheless the variation is not one that our investigation of Art. 232 will allow. For we take $\pi \int y^2 \varpi \, dy$ to denote the volume generated by the revolution of BA round the axis of x; and then we should be taking $\pi \int y^2 (\varpi + \delta\varpi) \, dy$ for the volume generated by the revolution of BD and DE: but it is obvious that the last expression really represents the volume generated by the revolution of DE, and so our expression would omit altogether the volume generated by the revolution of BD.

In order to allow for such a change of figure we ought to use for the general expression of the volume not $\pi \int y^2 \varpi \, dy$, but $2\pi\int (a-x)\,y\,dy$. Then we proceed to find a maximum or a minimum of $\int_0^{y_1}\left(\frac{y}{1+\varpi^2}+\lambda ay-\lambda xy\right)dy$, where λ is a constant.

This leads in the usual way to

$$-\lambda y+2\frac{d}{dy}\frac{y\varpi}{(1+\varpi^2)^2}=0;$$

therefore $$-\lambda y^2+\frac{4y\varpi}{(1+\varpi^2)^2}=\text{a constant},$$

and the constant must be zero.

But now we have an integrated term, namely $\frac{-2y\varpi\,\delta x}{(1+\varpi^2)^2}$. This vanishes at the lower limit, for which $y=0$; but does not vanish at the upper limit. For such a variation as the diagram represents we have δx negative at B, and thus we have a *positive* term in the variation of the integral. Hence when we say that we have a maximum in Art. 232, we must remember that such a variation as that illustrated in the present Article is excluded: such a variation in fact would *increase* the supposed maximum.

234. In Art. 221 we proposed to return again to the discontinuous solution in order to remove the arbitrary appearance which seemed to belong to it.

Let BC and CA be the two components of the discontinuous solution.

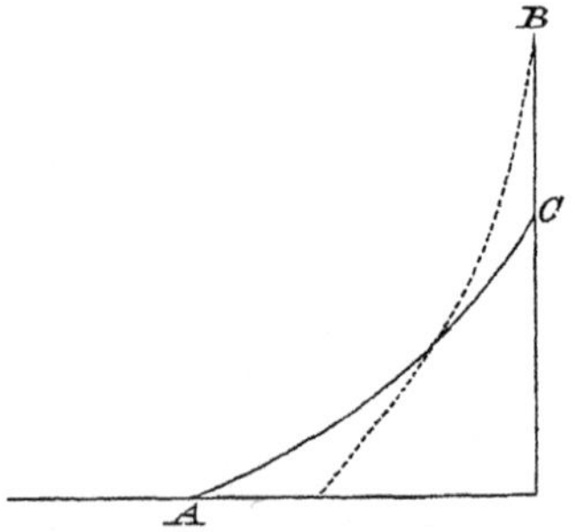

When we take x for the independent variable as in Art. 216 this solution is not very clearly suggested; the relation $p=\infty$

which belongs to the rectilinear part BC does not satisfy the fundamental equation unless we also make $\lambda = 0$, and this will not suit the curve part CA.

Although the discontinuous solution does not appear to be suggested by the fundamental equation, yet we may be naturally led to investigate it by the consideration that the given base constitutes a boundary which is not to be transgressed by our generating curve. We have already had examples of this character. Sometimes the boundary is suggested by the fundamental equation, and sometimes not: see Articles 68, 100, and 111.

But, by making x the independent variable, and ascribing variations to y only, we have in fact put it out of our power to recognise the discontinuous solution. Let the dotted curve represent a curve lying close to the discontinuous solution; then we cannot pass from one to the other by *infinitesimal* changes in y, and so the method we adopt is really unsuitable to the full discussion of the problem. Accordingly to bring the discontinuous solution into notice, while retaining x as the independent variable in the investigation, we have to modify our expression for the resistance; see Art. 222.

Now turn to the solution of Art. 232, in which y is made the independent variable.

Suppose, as there, that for the curve part we have

$$2\lambda y^2 = \frac{2y\varpi}{(1+\varpi^2)^2};$$

then the term of the first order in the variation vanishes so far as the part CA is concerned; and for the part CB it reduces to

$$-\int_0^{y_1} \frac{d}{dy}\left\{-\frac{2y\varpi}{(1+\varpi^2)^2} + 2\lambda y^2\right\}\delta x\, dy,$$

that is, since $\varpi = 0$, to

$$-4\int_0^{y_1} \lambda y\, \delta x\, dy.$$

As δx is necessarily negative, this is essentially positive, and so corresponds to a minimum. Hence we have simply another illustration and confirmation of the general theory laid down in Art. 17.

235. Suppose we add another condition to the problem stated in Art. 215 and restricted as in Art. 220: let the height be given, that is, the distance of the vertex from the base.

If the height of the continuous or discontinuous solution already found is less than the given height, the solution already found must still be adopted; and to produce the required height

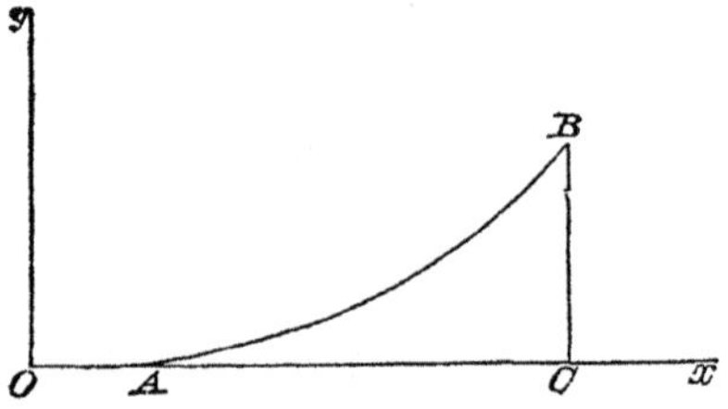

a portion of the axis of x must be used. Thus in the diagram, if OC is the given height, we must add the straight line OA to the curve part AB. It will be seen that $y = 0$ is a solution of the fundamental equation (1) of Art. 216.

If the given height be less than that of the continuous or discontinuous solution already obtained, we may still use a portion of the axis of x to fulfil the prescribed conditions in the manner of Art. 134, when the cone points inwards. Or if we understand the condition in another way, we must take as part of the boundary a straight line at right angles to the axis of x and proceed in the manner of Art. 190.

236. We may here notice the more general problem in which it is required to make $\int y\,\phi(p)\,dx$ a maximum or minimum while $\int \psi(y)\,dx$ is constant, supposing that $\psi(y)$ vanishes with y, that the value of x at the upper limit of integration is fixed, and that y is to vanish at the lower limit of integration.

We have to find a maximum or minimum of

$$\int_{x_0}^{a} \{y\,\phi(p) + \lambda\psi(y)\}\,dx,$$

where λ is a constant. Denote this by u; change x_0 into $x_0 + dx_0$ and vary y and p: then

$$\delta u = \int_{x_0+dx_0}^{a} \{(y+\delta y)\,\phi(p+\delta p) + \lambda\psi(y+\delta y)\}\,dx$$

$$-\int_{x_0}^{a} \{y\phi(p) + \lambda\psi(y)\}\,dx$$

$$= \int_{x_0}^{a} \{(y+\delta y)\,\phi(p+\delta p) + \lambda\psi(y+\delta y) - y\phi(p) - \lambda\psi(y)\}\,dx$$

$$-\int_{x_0}^{x_0+dx_0} \{(y+\delta y)\,\phi(p+\delta p) + \lambda\psi(y+\delta y)\}\,dx.$$

This is exact; now approximate to the second order: then the first of these integrals becomes in the usual way

$$y\phi'(p)\,\delta y + \int\left\{\phi(p) + \lambda\psi'(y) - \frac{d}{dx}[y\phi'(p)]\right\}\delta y\,dx$$

$$+\frac{1}{2}\int\{\lambda\psi''(y)\,(\delta y)^2 + 2\phi'(p)\,\delta y\,\delta p + y\phi''(p)\,(\delta p)^2\}\,dx,$$

all to be taken between the limits x_0 and a.

The second integral becomes

$$-\{y\phi(p) + \lambda\psi(y)\}_0\,dx_0$$

$$-\{\phi(p)\,\delta y + y\phi'(p)\,\delta p + \lambda\psi'(y)\,\delta y\}_0\,dx_0$$

$$-\frac{1}{2}\{p\phi(p) + yy''\phi'(p) + \lambda p\psi'(y)\}_0\,(dx_0)^2,$$

for we know that to the second order

$$\int_{x_0}^{x_0+dx_0} \chi(x)\,dx = \{\chi(x)\,dx + \frac{1}{2}\chi'(x)\,(dx)^2\}_0.$$

To make the term under the integral sign of the first order vanish, we require in the usual way

$$\phi(p) + \lambda\psi'(y) - \frac{d}{dx}[p\phi'(p)] = 0;$$

and this leads to

$$y\phi(p) + \lambda\psi(y) = yp\phi'(p) + \text{constant};$$

this constant must be zero, since zero is to be a value of y and $\psi(y)$ vanishes with y.

The terms of the first order outside the integral sign obviously vanish, except $y\phi'(p)\,\delta y$ at the upper limit: this term vanishes if y is constant at the upper limit, but if y is not constant there we must have $\phi'(p)$ zero at the upper limit.

Thus we are left with the following expression for the variation of u, which is of the second order:

$$\frac{1}{2}\int_{x_0}^{a} \{\lambda\psi''(y)(\delta y)^2 + 2\phi'(p)\,\delta y\,\delta p + y\phi''(p)(\delta p)^2\}\,dx$$

$$-\{\phi(p)\,\delta y + y\phi'(p)\,\delta p + \lambda\psi'(y)\,\delta y\}_0\,dx_0$$

$$-\frac{1}{2}\{p\phi(p) + yy''\phi'(p) + \lambda p\psi'(y)\}_0\,(dx_0)^2.$$

Now the following relation, true to the second order, exists between the differential and the variations

$$\delta y_0 = -\left\{pdx + \frac{1}{2}\frac{d^2y}{dx^2}(dx)^2 + \delta p\,dx\right\}_0.$$

See Todhunter's *History of the Calculus of Variations*, page 330, observing that the $\psi(x)$ there is now simply x. But we only require now the relation $\delta y_0 = -(pdx)_0$ which is true to the first order. Thus observing that $y_0 = 0$, we find that the part of the term of the second order in the variation which is outside the integral sign becomes

$$\frac{1}{2}\{p\phi(p) + \lambda p\psi'(y)\}_0\,(dx_0)^2.$$

The part of the term of the second order in δu which is under the integral sign may be transformed, as in Art. 26, to

$$\frac{1}{2}\{\phi'(p)(\delta y)^2\}_1 - \frac{1}{2}\{\phi'(p)(\delta y)^2\}_0$$

$$+\frac{1}{2}\int_{x_0}^{a}\{\phi''(p)[y(\delta p)^2 - y''(\delta y)^2] + \lambda\psi''(y)(\delta y)^2\}\,dx.$$

This may be modified in form by the aid of the following relation,

$$\lambda\psi'(y) - \frac{\lambda}{y}\psi(y) = yy''\phi''(p);$$

this follows by differentiating the equation

$$y\phi(p) + \lambda\psi(y) = yp\,\phi'(p).$$

237. As an example of the preceding general investigation, take $\phi(p) = p^2$ and $\psi(y) = y^3$; so that $\int yp^2 dx$ is to be a minimum, while $\int y^3 dx$ has a constant value. And let us suppose the value of y given at the upper limit. We have

$$y\phi(p) + \lambda\psi(y) = yp\phi'(p) + \text{constant} \quad \text{.........(1)};$$

that is, since the constant must be zero, $yp^2 + \lambda y^3 = 2yp^2$;

therefore $$p^2 = \lambda y^2 \quad \text{..............................(2)}.$$

For λ put γ^2; thus from (2) we get

$$\frac{dx}{dy} = \frac{1}{\gamma y};$$

therefore $y = Be^{\gamma x}$ where B is a constant.

The constants B and γ will have to be found from the known value of y at the upper limit, and the known value of $\int y^3 dx$.

In this example, supposing γ to be positive, we have y zero when $x = -\infty$; so that $x_0 = -\infty$.

The conditions for a minimum may be considered to be satisfied; as we see by taking the term of the second order in δu in the last form of the preceding Article.

But we do not thus get the *least* value of $\int yp^2\,dx$; for as in Art. 219 we can get as small a value as we please of this integral by a zigzag boundary: this boundary is suggested by the fact that the fundamental equation (1) is satisfied by a constant value of y.

Suppose that we impose another condition in the manner of Art. 235; let it be given that the curve is not to extend on the negative side of the axis of y. Then we must seek for a solution by combining a portion of the axis of y with the curve given by (2). Thus we obtain a figure composed of pieces like the OA and AP of the diagram of Art. 201. The solution must be taken as in former examples to be a *limit* towards which we approach by drawing curves close to the discontinuous boundary. The curves must of course be supposed so taken as to make $\int yp^2\,dx$ finite in the neighbourhood of the origin; for example, this will be secured if near the origin y varies as $\sqrt{x}$.

238. The problem of Art. 236 may also be treated by taking y as the independent variable. Put ϖ for $\frac{dx}{dy}$, and let $\phi(p)\,dx$ transform into $f(\varpi)\,dy$; then we have to find the maximum or minimum of

$$\int_0^{y_1} \left\{ yf(\varpi) + \lambda\psi(y)\,\varpi \right\} dy.$$

This leads in the usual way to

$$yf'(\varpi) + \lambda\psi(y) = 0 \quad \text{.....................(1)};$$

and if y_1 is susceptible of an increment we must also have

$$\left\{ yf(\varpi) + \lambda\psi(y)\,\varpi \right\}_1 = 0 \quad \text{.....................(2)}.$$

Then we have left a variation of the second order consisting of

$$\frac{1}{2}\int_0^{y_1} yf''(\varpi)(\delta\varpi)^2\,dx,$$

together with the part of the second order in

$$\int_{y_1}^{y_1+dy_1}\left\{yf(\varpi+\delta\varpi)+\lambda\psi(y)(\varpi+\delta\varpi)\right\}dy;$$

the latter part is

$$\left\{yf'(\varpi)+\lambda\psi(y)\right\}_1\delta\varpi_1\,dy_1$$

$$+\frac{1}{2}\left\{f(\varpi)+yf'(\varpi)\frac{d\varpi}{dy}+\lambda\psi'(y)\,\varpi+\lambda\psi(y)\frac{d\varpi}{dy}\right\}_1(dy_1)^2,$$

which by means of (1) reduces to

$$\frac{1}{2}\left\{f(\varpi)+\lambda\psi'(y)\,\varpi\right\}_1(dy_1)^2.$$

This may be modified in form by the aid of (2).

CHAPTER XI.

JAMES BERNOULLI'S PROBLEM.

239. A and B are fixed points; a curve of given length is to be drawn from A to B, having the following property:

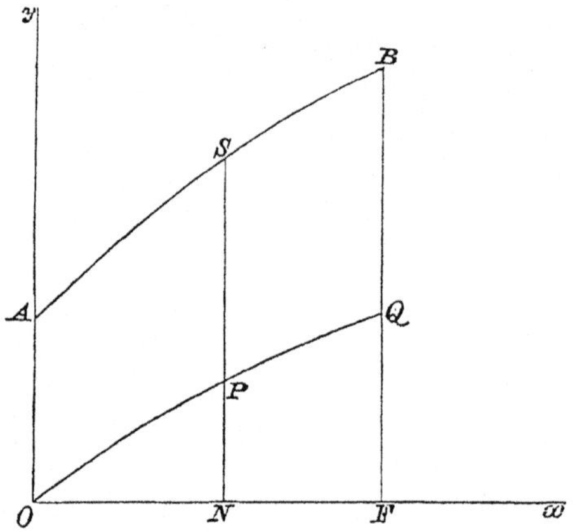

at any point S of the curve draw SN perpendicular to the fixed straight line OF, and take the ordinate NP equal to the arc AS; then the curve traced out by P is to enclose a maximum or minimum area.

The problem is a particular case of one which was given by James Bernoulli: see Todhunter's *History of the Calculus of Variations*, page 453.

240. It will be apparent on a little reflection that there must be both a *greatest* and a *least* value of the area bounded by OPQ. The fixed straight line OF will be taken for the axis of x, and the axis of y will be taken to pass through A. We may remark that by the nature of the curve OPQ, the tangent at any point P cannot be inclined to the axis of x at an angle less than 45°.

We assume that the curve ASB is to be comprised between the ordinates at A and B.

Let $$OA = h, \quad OF = a, \quad FB = k.$$

Let x and y be the co-ordinates of S, and let s denote the length of the arc AS: then

$$s = \int_0^x \sqrt{(1+p^2)}\, dx,$$

and we require that $\int_0^a s\, dx$ shall be a maximum or minimum, while $\int_0^a \sqrt{(1+p^2)}\, dx$ is constant.

Let $u = \int_0^a \{s + \lambda \sqrt{(1+p^2)}\}\, dx$ where λ is a constant. Change p to $p + \delta p$; then to the second order

$$\delta u = \int_0^a \left\{\delta s + \frac{\lambda p\, \delta p}{\sqrt{(1+p^2)}} + \frac{\lambda (\delta p)^2}{2(1+p^2)^{\frac{3}{2}}}\right\} dx.$$

Also $$\delta s = \int_0^x \frac{p \delta p\, dx}{\sqrt{(1+p^2)}} + \frac{1}{2}\int_0^x \frac{(\delta p)^2\, dx}{(1+p^2)^{\frac{3}{2}}}.$$

And $\int \delta s\, dx = x\delta s - \int x \frac{d\delta s}{dx}\, dx$, so that

$$\int_0^a \delta s\, dx = a \int_0^a \left\{\frac{p\delta p}{\sqrt{(1+p^2)}} + \frac{(\delta p)^2}{2(1+p^2)^{\frac{3}{2}}}\right\} dx$$

$$- \int_0^a x \left\{\frac{p\delta p}{\sqrt{(1+p^2)}} + \frac{(\delta p)^2}{2(1+p^2)^{\frac{3}{2}}}\right\} dx$$

$$= \int_0^a (a - x) \left\{\frac{p\delta p}{\sqrt{(1+p^2)}} + \frac{(\delta p)^2}{2(1+p^2)^{\frac{3}{2}}}\right\} dx.$$

Thus, finally,

$$\delta u = \int_0^a (\lambda + a - x) \left\{ \frac{p\delta p}{\sqrt{(1+p^2)}} + \frac{(\delta p)^2}{2(1+p^2)^{\frac{3}{2}}} \right\} dx.$$

From the value of δu we see that our problem coincides with the following: find a curve of given length between the fixed points A and B for which $\int (a-x)\, ds$ is a maximum or a minimum. Hence, we require in effect a curve of given length which shall have its centre of gravity at a maximum or minimum distance from the straight line $x=a$. The curve is well known to be a catenary.

The term in δu which is of the first order must now be transformed in the usual way: we have

$$\int (\lambda + a - x) \frac{p\delta p}{\sqrt{(1+p^2)}}\, dx = \frac{(\lambda+a-x)\, p\delta y}{\sqrt{(1+p^2)}} - \int \delta y \frac{d}{dx} \frac{(\lambda+a-x)p}{\sqrt{(1+p^2)}}\, dx.$$

When this is taken between limits the part outside the integral sign vanishes; to make the other part vanish we must put

$$\frac{(\lambda+a-x)\,p}{\sqrt{(1+p^2)}} = \text{a constant, say } c \;\ldots\ldots\ldots\ldots(1);$$

therefore

$$\frac{1}{p^2} = \frac{(\lambda+a-x)^2 - c^2}{c^2} \;\ldots\ldots\ldots\ldots\ldots\ldots\ldots(2).$$

Thus δu reduces to the term of the second order

$$\frac{1}{2} \int_0^a \frac{(\lambda + a - x)(\delta p)^2}{(1+p^2)^{\frac{3}{2}}}\, dx.$$

241. The curve determined by (2) is a catenary having its directrix parallel to the axis of y; and c is *numerically* equal to the parameter, which, in the language of Statics, measures the tension at the lowest point.

By putting $x=a$ in (2) we see that λ is *numerically* equal to the perpendicular distance of B from the directrix of the catenary. But we shall have to pay careful attention to the signs of c and λ.

From (1) we have

$$\lambda + a - x = \frac{c\sqrt{(1+p^2)}}{p};$$

therefore $$1 = \frac{c}{p^2\sqrt{(1+p^2)}}\frac{dp}{dx} \quad\text{.....................(3)};$$

thus c is of the same sign as $\frac{dp}{dx}$, so that c is positive or negative according as the required curve is convex or concave to the axis of x.

242. We will now assume than h is less than k; from our discussion of this case it will be easy to see how to proceed when h is not less than k.

If then the curve ASB is concave to the axis of x, we have c negative; and then $\lambda + a - x$ is negative by (1); thus λ is negative. And since $\lambda + a - x$ is negative, δu is a negative quantity of the second order, and we have a maximum.

If the curve ASB is convex to the axis of x we have c positive; and then $\lambda + a - x$ is positive by (1), and so λ is positive. And since $\lambda + a - x$ is positive, δu is a positive quantity of the second order, and we have a minimum.

243. Now let us consider how far these solutions are really admissible. Take for instance the maximum. Begin with a given length very little greater than the straight line which would join A to B; it is obvious from statical considerations that the required catenary exists. Suppose the given length gradually increased; then it is obvious in the same way that the required catenary always exists until we arrive at the case in which the catenary *touches* OA at A. If the given length be still further increased, the catenary would cut the axis of y above A, and the solution is no longer tenable.

244. Now, as by supposition we are confined by the axis of y, we are, as in former problems, led to enquire whether the solution will not be composed in some cases of part of the axis of y, together with a curve; see Arts. 221 and 234.

Suppose then, if possible, that the required line consists of a straight line of the length $y_0 - h$ measured from A along the positive direction of the axis of y, and a curve proceeding from the point in the axis of y which has y_0 for ordinate to B.

Thus we now ask for a maximum of

$$\int_0^a (y_0 - h + s)\, dx,$$

where
$$s = \int_0^x \sqrt{(1+p^2)}\, dx;$$

and the given length must be equal to

$$y_0 - h + \int_0^a \sqrt{(1+p^2)}\, dx.$$

Therefore we have now to find a maximum of

$$\int_0^a \{y_0 - h + s + \lambda \sqrt{(1+p^2)}\}\, dx + \lambda (y_0 - h),$$

that is, of

$$\int_0^a \{s + \lambda \sqrt{(1+p^2)}\}\, dx + (\lambda + a)(y_0 - h).$$

The only difference between the present form of the problem and that of Art. 240 arises from the presence now of terms in the variation outside the integral sign. We have $(\lambda + a)\,\delta y_0$ from $(\lambda + a)(y_0 - h)$; and from the integral itself we obtain $-c\delta y_0$: thus, on the whole, we have $(\lambda + a - c)\,\delta y_0$.

Hence, as this must vanish, the following relation must hold,

$$\lambda + a - c = 0 \quad\ldots\ldots\ldots\ldots\ldots\ldots(4).$$

Now, the curve being supposed concave to the axis of x, we know that λ is negative, and $\lambda + a$ is numerically equal to the distance of A from the directrix of the catenary: hence it follows from (4) that the catenary must *touch* the axis of y at the point corresponding to y_0.

Thus the conditions required for a maximum are all satisfied in this solution.

This solution holds only so long as y_0 is not greater than k, the limiting case being that in which $y_0 = k$, and the catenary degenerates into a straight line: the given length is then equal to $k - h + a$.

245. It should be remarked that the solution of the preceding Article is suggested by the general equation (1) of Art. 240. For if we suppose $x = 0$ so that p is infinite and put $\lambda + a = c$, that equation is satisfied. And the relation $\lambda + a = c$ holds, as we have seen, for the curve part of the solution also.

Thus the discontinuous solution arises in fact from combining two solutions which are both involved in the ordinary general result of the Calculus of Variations.

In Art. 234 we had to account for the fact that the discontinuous solution did not appear to be very obviously contained in the general result; whereas in the present problem the discontinuous solution is so contained. The difference perhaps depends on the fact that here, as we see in Art. 240, we do not imperatively require that δy should be infinitesimal in our investigations, for y does not explicitly occur under the integral sign: we merely change p into $p + \delta p$, and our process requires that δp should be indefinitely small in comparison with p, so as to allow us to expand $\sqrt{1 + (p + \delta p)^2}$ suitably.

246. We may observe that for a given length of string there is only one solution which furnishes a maximum; namely, either the continuous solution of Art. 242, or the discontinuous solution of Art. 244, according as the given length is less or greater than a certain value. This will be sufficiently obvious from statical considerations. Suppose a fixed point A on a smooth horizontal table, and a fixed point B above it; let a heavy uniform string have its ends fastened at A and B; and suppose the length of this string to be not less than that of the straight line which joins A to B, and not greater than the $k - h + a$ of Art. 244. Then we may admit that this string will be in equilibrium in one position and only in one: if the length is below a certain value no part of the string will be in contact with the table, and if the length is

above this value, part of the string will be in contact with the table. If l denote the length of the string when the string touches the table at A, we find from the nature of the catenary that

$$a = \frac{l^2 - \gamma^2}{2\gamma} \log \frac{l + \gamma}{l - \gamma},$$

where γ is put for $k - h$.

Of course these statements may be demonstrated by analysis without having recourse to statical considerations. When the length of a string and the positions of its extreme points are given, we can form equations for determining a catenary with its base horizontal fulfilling the given conditions: we find that there is only one value of the parameter c, and the greater the given length is the less is this parameter.

For let l denote the given length of a string; let h denote the horizontal distance of the fixed points; and let b denote the difference of the distances of the fixed points from a given horizontal line. Then it is shewn in works on Statics that the parameter c of the required catenary is determined by the equation

$$l^2 - b^2 = c^2 (e^{\frac{h}{c}} + e^{-\frac{h}{c}} - 2).$$

Expand the right-hand member in powers of c; thus we obtain

$$l^2 - b^2 = 2h^2 \left\{ \frac{1}{\underline{|2}} + \frac{1}{\underline{|4}} \frac{h^2}{c^2} + \frac{1}{\underline{|6}} \frac{h^4}{c^4} + \dots \right\}.$$

The expression on the right-hand side decreases continually as c increases; and thus we see that the equation will give only one value of c^2 corresponding to an assigned value of l, and the greater l is the smaller c^2 is.

Hence two catenaries with parallel bases cannot intersect in more than two points; for if they could each would have a less parameter than the other, which is absurd. Hence as the given length is gradually increased from the least admissible value, a series of catenaries is obtained each lying entirely outside the preceding, until we arrive at the length denoted by l in the former part of this Article.

Then after passing this length we can obtain another series of lines each consisting of a straight line and a catenary. Now consider two members of this series. If the catenary corresponding to that which has the longer straight line be continued upwards from its lowest point, it will obviously cut the other catenary; and as the two catenaries have also the fixed point in common they cannot cut in any third point. Hence of the two members of the series that which has the longer straight line is the longer; for it is *outside* the other except where the two coincide. Hence it follows that corresponding to a given length there is only one member of the series. Thus as there is only one solution for a given length of string which furnishes a *maximum*, and as we are sure there must be a *greatest* value, we may infer that the maximum value is the greatest value.

247. Let us now return to that point of the investigation which was reached at the end of Art. 244; and suppose that the given length is greater than $k - h + a$.

We shall find that there is now a maximum when we take a straight line of length $y_0 - h$ along the axis of y from A, and a catenary *convex* to the axis of x touching the axis of y at the point corresponding to y_0, and passing through B.

The investigation is similar to that of Art. 244; now we have c positive by equation (3), and as p is negative we see that $\lambda + a - x$ is negative by equation (1). As in Art. 244 we arrive at the condition $\lambda + a - c = 0$, which makes the catenary touch the axis of y at the point corresponding to y_0. But although we thus obtain a *maximum*, we do not obtain the *greatest* value of the area; for there is another solution which is now admissible.

248. We see in fact that $p = 0$ is *a* solution of (1) provided of course that $c = 0$; and we will now interpret this in combination with $p = \infty$, which is also a solution of (1) supposing that $\lambda + a - x = 0$ since $c = 0$.

Draw BC parallel to the axis of x; take a point D on the axis of y such that $AD + DC + CB =$ the given length; then we may suppose that our required curve is made up of AD, DC and CB. Corresponding to the curve OPQ of Art. 239 we have now the straight line RPQ where $OR = AD + DC$.

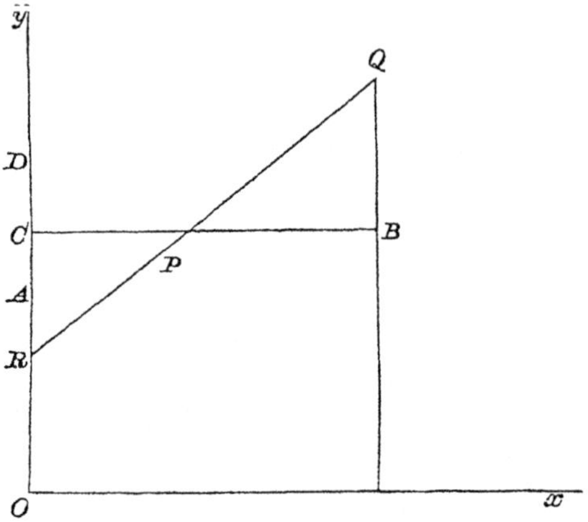

This solution might be treated in the manner of Art. 244. We now ask for a maximum of

$$\int_0^a (y_0 - h + y_0 - k + s)\,dx,$$

where $$s = \int_0^x \sqrt{(1+p^2)}\,dx;$$

and the given length must be equal to

$$y_0 - h + y_0 - k + \int_0^a \sqrt{(1+p^2)}\,dx.$$

Hence proceeding as before we arrive at the equation

$$\frac{(\lambda + a - x)\,p}{\sqrt{(1+p^2)}} = \text{a constant } c,$$

and we have as the term in the variation outside the integral sign

$$\{2(a+\lambda) - c\}\,\delta y_0.$$

Hence we obtain a solution by the suppositions

$$p = 0, \quad c = 0, \quad a + \lambda = 0.$$

249. With regard to this solution I remark

I. It is certain that we obtain the *greatest* possible area in this way. For the ordinate of Q the extreme point of the derived line is of course equal to the whole given length; and the derived line becomes in this case a straight line inclined at an angle of 45° to the axis, and so the ordinates of this derived curve starting from Q diminish *more slowly* than for any other possible form of the primary curve: see Art. 240.

II. This solution, like the solutions for the cases already considered, is fairly deduced by the Calculus of Variations.

III. Should any person object that the solution does not strictly apply to the problem, for we were required to draw a *curve* from A to B, the answer must be similar to remarks already made. The proposed solution must be regarded as a *limit*. We may conceive a curve drawn from A to B, first running upwards close to the axis of y, then turning sharply and descending close to the ascending part until it arrives at about the level of B, then turning off towards B in a direction nearly parallel to the axis of x. Such a curve will give us for the area of the derived curve a result falling only infinitesimally short of what we have shewn to be the greatest possible value.

250. We will now briefly state the result with respect to a *minimum*.

If the given length be very little greater than the straight line which would join A to B, we are certain that a catenary can be drawn from A to B convex to the axis of x. This will give a minimum. The solution will hold up to the limiting case in which the catenary *touches* the ordinate of B at B.

If the given length be greater than that which corresponds to this limiting case the required line is made up of a portion $k - y_1$ of the ordinate at B measured from B towards the axis of x, and a catenary which touches the ordinate of B at the point corresponding to y_1. This will give a minimum. The solution will hold as long as y_1 is greater than h.

In the cases hitherto considered there is only one solution for a given length; and the solution is not only a *minimum*, but corresponds to the *least* area of the derived curve.

If the given length is greater than $k - h + a$ we have a choice of two solutions. One consists of a portion $k - y_1$ of the ordinate of B and a catenary *concave* to the axis of x, and touching the ordinate of B at the point corresponding to y_1 which is less than h. This solution gives a *minimum*. The other solution consists of the straight line $y = h$ together with such portions of the ordinate at B as may be required in addition to produce the given length. This solution corresponds to the *least* area of the derived curve, as we see by considerations like those already employed.

251. It is easy to give other examples like that of Art. 248. Return to the diagram of Art. 239, and suppose we require not that the area $OPQF$ shall be a maximum or a minimum, but that the volume generated by the revolution of this area round Ox shall be a maximum or a minimum.

Retaining the same notation as before, since the volume will be $\pi \int s^2\, dx$, we have to find a maximum or minimum of

$$\int_0^a \left\{ s^2 + 2\lambda \sqrt{(1+p^2)} \right\} dx.$$

Denote this by u; then to the second order

$$\delta u = 2\int_0^a \left\{ s\delta s + \frac{(\delta s)^2}{2} + \frac{\lambda p\, \delta p}{\sqrt{(1+p^2)}} + \frac{\lambda\, (\delta p)^2}{2\, (1+p^2)^{\frac{3}{2}}} \right\} dx.$$

And $$\int s\delta s\, dx = v\delta s - \int v \frac{d\delta s}{dx}\, dx,$$

where v stands for $\int_0^x s dx$; so that

$$\int_0^a s\delta s\, dx = V \int_0^a \left\{ \frac{p\delta p}{\sqrt{(1+p^2)}} + \frac{(\delta p)^2}{2\, (1+p^2)^{\frac{3}{2}}} \right\} dx$$

$$-\int_0^a v\left\{\frac{p\delta p}{\sqrt{(1+p^2)}}+\frac{(\delta p)^2}{2(1+p^2)^{\frac{3}{2}}}\right\}dx,$$

where $V=\int_0^a s dx$. Hence

$$\delta u = 2\int_0^a (\lambda+V-v)\left\{\frac{p\delta p}{\sqrt{(1+p^2)}}+\frac{(\delta p)^2}{2(1+p^2)^{\frac{3}{2}}}\right\}dx+\int_0^a(\delta s)^2 dx.$$

Hence we obtain in the usual way

$$\frac{(\lambda+V-v)p}{\sqrt{(1+p^2)}} = \text{a constant} = c \text{ say} \quad\ldots\ldots\ldots\ldots (5),$$

and then δu reduces to $\int_0^a (\lambda+V-v)\,\frac{(\delta p)^2}{(1+p^2)^{\frac{3}{2}}}dx+\int_0^a(\delta s)^2 dx.$

From (5) we have

$$\lambda+V-v=\frac{c\sqrt{(1+p^2)}}{p}=\frac{c\dfrac{ds}{dx}}{\sqrt{\left\{\left(\dfrac{ds}{dx}\right)^2-1\right\}}};$$

differentiate with respect to x; thus

$$s=\frac{c\dfrac{d^2s}{dx^2}}{\left\{\left(\dfrac{ds}{dx}\right)^2-1\right\}^{\frac{3}{2}}}.$$

Multiply by $\dfrac{ds}{dx}$ and integrate; thus

$$\frac{s^2}{2}=-\frac{c}{\left\{\left(\dfrac{ds}{dx}\right)^2-1\right\}^{\frac{1}{2}}}+\text{constant}\ldots\ldots\ldots\ldots\ldots(6).$$

The equations (5) and (6) are not simple enough to furnish us with the relation between x and y; but for our purpose the most interesting point is that if the given length be large enough, we can solve (5) by the combination of $p=0$ with $p=\infty$ in the manner of Art. 248.

We first take $x=0$ which gives $v=0$ and $p=\infty$; also we make $\lambda+V=0$ so that $c=0$; then we take $p=0$. Thus

δu becomes $-\int \frac{v\,(\delta p)^2\,dx}{(1+p^2)^{\frac{3}{2}}}$, which is necessarily negative; for $\int_0^a (\delta s)^2\,dx$ vanishes since $p=0$: so that we have a maximum. And by the reasoning of Art. 249 we are sure that this is not only a *maximum*, but that it is the *greatest* possible result.

252. Let us propose a problem in polar co-ordinates like that of Art. 239. An arc ASB of given length is to join the fixed points A and B; on OS take OP equal to *the square root* of AS: required the curve ASB so that the polar area generated by OP may be a maximum or a minimum.

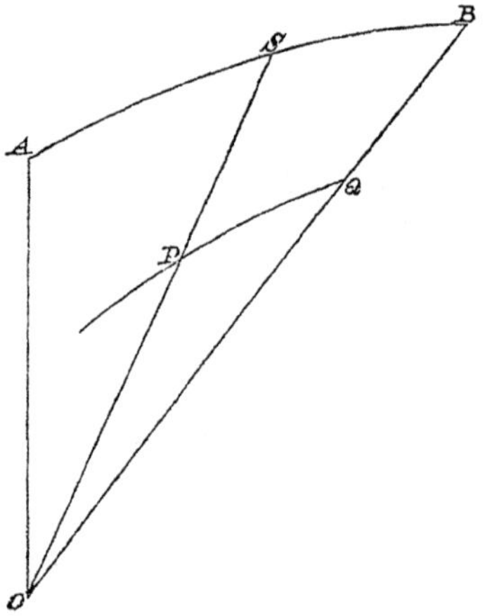

The area generated by $OP = \frac{1}{2}\int (OP)^2\,d\theta$, and therefore it varies as $\int s\,d\theta$ where $AS = s$. Let $u = \int_0^a \{s + \lambda\sqrt{(r^2+p^2)}\}\,d\theta$, where λ is a constant, $OS = r$, and p stands for $\frac{dr}{d\theta}$; also $s = \int_0^\alpha \sqrt{(r^2+p^2)}\,d\theta$.

Proceed as in Arts. 239 and 251; thus we shall find that to the first order

$$\delta u = \int_0^\alpha \frac{(\lambda + \alpha - \theta)(r\delta r + p\delta p)\,d\theta}{\sqrt{(r^2+p^2)}}.$$

This must be made to vanish in the ordinary way by the relation

$$\frac{(\lambda+\alpha-\theta)\,r}{\sqrt{(r^2+p^2)}}-\frac{d}{d\theta}\,\frac{(\lambda+\alpha-\theta)\,p}{\sqrt{(r^2+p^2)}}=0 \;\ldots\ldots\ldots\ldots (7).$$

Then the term in δu of the second order is

$$\int_0^\alpha \frac{(\lambda+\alpha-\theta)\,(p\delta r-r\delta p)^2\,d\theta}{2\,(r^2+p^2)^{\frac{3}{2}}}\,;$$

thus $\lambda+\alpha-\theta$ must be always negative for a maximum and always positive for a minimum.

Although we cannot deduce from (7) the explicit relation between r and θ, yet we shall see that if the given length be large enough we must have discontinuity of the kind exemplified in Arts. 244, 247 or 248.

Equation (7) when developed becomes

$$r\,(\theta-\lambda-\alpha)\left\{r^2+2\left(\frac{dr}{d\theta}\right)^2-r\,\frac{d^2r}{d\theta^2}\right\}=\frac{dr}{d\theta}\left\{r^2+\left(\frac{dr}{d\theta}\right)^2\right\}\ldots\ldots (8)\,;$$

hence we can infer that r does not admit of a maximum or a minimum at any point between A and B. For if $\frac{dr}{d\theta}$ could change sign $r^2+2\left(\frac{dr}{d\theta}\right)^2-r\,\frac{d^2r}{d\theta^2}$ must change sign at the same point; that is, there would be a change as to concavity and convexity at the point where r is a maximum or minimum: and this is impossible.

Hence we conclude that equation (7) will not supply curves of unlimited length between A and B; so that when the given length is large enough we must seek for a discontinuous solution.

If we proceed as in Art. 244 we shall have outside the integral sign in δu the term

$$\left\{\lambda+\alpha-\frac{(\lambda+\alpha-\theta)\,p}{\sqrt{(r^2+p^2)}}\right\}_0\delta r_0,$$

that is

$$(\lambda+\alpha)\left\{1-\frac{p}{\sqrt{(r^2+p^2)}}\right\}_0\delta r_0.$$

To make this vanish we must either have $\lambda + \alpha = 0$ or p_0 infinite.

If we take p_0 infinite we have a curve touching the initial radius OA at the point where the curve leaves the radius; and thus the solution resembles those in Arts. 244 and 247.

If we take $\lambda + \alpha = 0$ we see from (8) that $p_0 = 0$; in this case our solution resembles that of Art. 248. If we start as in Art. 248 we have for the term in δu outside the integral sign

$$(\lambda + \alpha)\left\{2 - \frac{p}{\sqrt{(r^2 + p^2)}}\right\}_0 \delta r_0,$$

and to make this vanish we must have $\lambda + \alpha = 0$.

253. We might approximate to the solution of equation (8) in particular cases. For example if we take $\lambda + \alpha = 0$, then we shall find that for small values of θ we have approximately

$$r = b\left\{1 + \frac{\theta^2}{4} + \frac{5\theta^4}{128}\right\},$$

where b is the value of r corresponding to $\theta = 0$.

254. The term of the second order in δu of Art. 252 will vanish if $p\delta r - r\delta p$ is always zero; and thus it might seem as if we should not be sure of having obtained a maximum or a minimum. But this relation leads to $\delta r = Cr$ where C is a constant; and it will be found that this is inconsistent with our condition that the length of the curve is given.

255. Let p stand for $\frac{dz}{dx}$ and q for $\frac{dz}{dy}$: required a maximum or minimum of

$$\int_0^a \int_0^b S\,dx\,dy,$$

while

$$\int_0^a \int_0^b \sqrt{(1 + p^2 + q^2)}\,dx\,dy$$

is constant, and

$$S = \int_0^x \int_0^y \sqrt{(1 + p^2 + q^2)}\,dx\,dy.$$

This is an obvious extension of the problem of Art. 239 to space of three dimensions.

By the usual theory we must seek for a maximum or minimum of

$$\int_0^a \int_0^b \{S + \lambda \sqrt{(1 + p^2 + q^2)}\}\, dx\, dy,$$

where λ is a constant. Denote this by u, then to the first order

$$\delta u = \int_0^a \int_0^b \left\{\delta S + \frac{\lambda (p\delta p + q\delta q)}{\sqrt{(1 + p^2 + q^2)}}\right\} dx\, dy.$$

Now
$$\int \delta S\, dy = y \delta S - \int y \frac{d\delta S}{dy}\, dy\,;$$

therefore
$$\int \left\{\int \delta S\, dy\right\} dx = xy\, \delta S - y \int x \frac{d\delta S}{dx}\, dx$$
$$- x \int y \frac{d\delta S}{dy}\, dy + \iint xy \frac{d^2 \delta S}{dx\, dy}\, dx\, dy:$$

thus
$$\int_0^a \int_0^b \delta S\, dx\, dy = ab \int_0^a \int_0^b \omega\, dx\, dy$$
$$- b \int_0^a \int_0^b x\, \omega\, dx\, dy - a \int_0^a \int_0^b y\, \omega\, dx\, dy + \int_0^a \int_0^b xy\, \omega\, dx\, dy$$
$$= \int_0^a \int_0^b (a - x)(b - y)\, \omega\, dx\, dy,$$

where ω stands for $\dfrac{p\delta p + q\delta q}{\sqrt{(1 + p^2 + q^2)}}$.

Thus finally

$$\delta u = \int_0^a \int_0^b \frac{\{(x - a)(y - b) + \lambda\}(p\delta p + q\delta q)}{\sqrt{(1 + p^2 + q^2)}}\, dx\, dy.$$

From the value of δu we see that our problem coincides with the following: find a surface of given area within assigned limits such that the sum of the product of every element into $(x - a)(y - b)$ shall be a maximum or a minimum.

We will suppose that the surface is to pass through two given points; one on the axis of z, and the other on the straight line $x = a$ and $y = b$: we will call the former point A and the latter point B.

We transform δu by the usual method; and thus we find that for a maximum or a minimum we must have

$$\frac{d}{dx}\frac{(x-a)(y-b)+\lambda}{\sqrt{(1+p^2+q^2)}}p+\frac{d}{dy}\frac{(x-a)(y-b)+\lambda}{\sqrt{(1+p^2+q^2)}}q=0 \;\ldots\; (9).$$

There are also certain terms of the first order in δu which are outside the signs of *double* integration; to make these vanish, we shall require $q = 0$ when $y = 0$ and when $y = b$ for all values of x, and $p = 0$ when $x = 0$ and when $x = a$ for all values of y.

The term of the second order in δu will be found to be

$$\frac{1}{2}\int_0^a\int_0^b \frac{(x-a)(y-b)+\lambda}{(1+p^2+q^2)^{\frac{3}{2}}}\{(\delta p)^2+(\delta q)^2+(q\delta p-p\delta q)^2\}\,dx\,dy.$$

I do not propose to make use of the equation (9) in order to obtain a general solution, but only to shew that a certain discontinuous solution is admissible. Suppose the given area to be only infinitesimally greater than ab; then we may avail ourselves of the solution corresponding to $p = 0$ and $q = 0$. That is, we may take a plane parallel to the plane of (x, y), and make it pass through one of the fixed points, say B. Then this plane must be supposed to be connected with the fixed point A by a portion of the axis of z, or by a part of a very slender conical surface close to the axis of z. If the given area exceeds ab by a finite quantity, we can suppose that the surface rises from the fixed point A close to the axis of z to a sufficiently great height, and then descends again to the plane which passes through B, and is parallel to the plane of (x, y).

Thus, at any point of this plane we may suppose that

$$S = xy + S_1,$$

where S_1 is such that $S_1 + ab$ is equal to the given area; and the amount S_1 of area must be supposed to be placed close round the axis of z.

256. Let us now pay some attention to the more general problem given by James Bernoulli, of which that in Art. 239 is a particular case, and that in Art. 251 another.

Suppose then that the ordinate PN of the derived curve is to be a given function of the arc AS; say that

$$PN = \phi(s).$$

We then proceed to seek for a maximum or minimum of

$$\int_0^a \{\phi(s) + \lambda \sqrt{(1+p^2)}\}\, dx;$$

call it u; then to the second order

$$\delta u = \int_0^a \left\{\phi'(s)\,\delta s + \frac{1}{2}\phi''(s)\,(\delta s)^2 + \frac{\lambda p\,\delta p}{\sqrt{(1+p^2)}} + \frac{\lambda\,(\delta p)^2}{2\,(1+p^2)^{\frac{3}{2}}}\right\} dx,$$

and
$$\delta s = \int_0^x \frac{p\,\delta p\, dx}{\sqrt{(1+p^2)}} + \frac{1}{2}\int_0^x \frac{(\delta p)^2\, dx}{(1+p^2)^{\frac{3}{2}}}.$$

Now put $\psi(x)$ for $\int_0^x \phi'(s)\, dx$: then

$$\int \phi'(s)\,\delta s\, dx = \psi(x)\,\delta s - \int \psi(x)\,\frac{d\delta s}{dx}\, dx,$$

so that

$$\int_0^a \phi'(s)\,\delta s\, dx = \int_0^a \{\psi(a) - \psi(x)\} \left\{\frac{p\delta p}{\sqrt{(1+p^2)}} + \frac{(\delta p)^2}{2\,(1+p^2)^{\frac{3}{2}}}\right\} dx.$$

Hence
$$\delta u = \int_0^a \{\psi(a) - \psi(x) + \lambda\} \frac{p\delta p}{\sqrt{(1+p^2)}}\, dx$$

$$+ \frac{1}{2}\int_0^a \left\{[\psi(a) - \psi(x) + \lambda]\, \frac{(\delta p)^2}{(1+p^2)^{\frac{3}{2}}} + \phi''(s)\,(\delta s)^2\right\} dx.$$

To make the term in δu which is of the first order vanish, we put

$$\frac{\{\psi(a)-\psi(x)+\lambda\}\,p}{\sqrt{(1+p^2)}}=c \text{ a constant } \ldots\ldots\ldots (10).$$

As in the special case of $\phi(s)=s$, which we have already discussed, we see that there may be particular admissible solutions of (10) given by $p=0$ and $p=\infty$, as well as the general solution. We may put (10) in the form

$$\psi(a)-\psi(x)+\lambda=\frac{c\sqrt{(1+p^2)}}{p} \ldots\ldots\ldots\ldots\ldots (11);$$

therefore, by differentiation,

$$-\phi'(s)=-\frac{c}{p^2\sqrt{(1+p^2)}}\frac{dp}{dx},$$

so that

$$\phi'(s)\frac{ds}{dx}=\frac{c}{p^2}\frac{dp}{dx};$$

therefore

$$\phi(s)=-\frac{c}{p}+c_1 \ldots\ldots\ldots\ldots\ldots\ldots\ldots\ldots\ldots (12).$$

The term of the second order in δu

$$=\frac{1}{2}\int_0^a\left\{\frac{\psi(a)-\psi(x)+\lambda}{(1+p^2)^{\frac{3}{2}}}\,\frac{1+p^2}{p^2}\left(\frac{d\delta s}{dx}\right)^2+\phi''(s)\,(\delta s)^2\right\}dx;$$

taking the general solution (11), this becomes

$$\frac{1}{2}\int_0^a\left\{\frac{c}{p^3}\left(\frac{d\delta s}{dx}\right)^2+\phi''(s)\,(\delta s)^2\right\}dx.$$

If $\frac{c}{p^3}$ is of the same sign as $\phi''(s)$ this term will require no transformation; if $\frac{c}{p^3}$ is not of the same sign as $\phi''(s)$ the following transformation may be useful:

$$\int\frac{c}{p^3}\left(\frac{d\delta s}{dx}\right)^2dx=\frac{c}{p^3}\frac{d\delta s}{dx}\,\delta s-\int\delta s\,\frac{d}{dx}\left(\frac{c}{p^3}\frac{d\delta s}{dx}\right)dx;$$

and as $\delta s = 0$ at the limits, we get

$$\int_0^a \frac{c}{p^3}\left(\frac{d\delta s}{dx}\right)^2 dx = -\int_0^a \delta s \frac{d}{dx}\left(\frac{c}{p^3}\frac{d\delta s}{dx}\right) dx,$$

so that we have

$$\delta u = \frac{1}{2}\int_0^a \delta s \left\{\phi''(s)\,\delta s - \frac{d}{dx}\left(\frac{c}{p^3}\frac{d\delta s}{dx}\right)\right\} dx.$$

257. From equation (12) we obtain

$$\frac{dy}{dx} = \frac{c}{c_1 - \phi(s)};$$

therefore
$$\frac{dy}{ds} = \frac{c}{\sqrt{[c^2 + \{c_1 - \phi(s)\}^2]}},$$

and
$$\frac{dx}{ds} = \frac{c_1 - \phi(s)}{\sqrt{[c^2 + \{c_1 - \phi(s)\}^2]}}.$$

From these last two equations we must find x and y in terms of s by integration. Thus from the last we obtain $x + c_2$ as a function of s, c and c_1, where c_2 is another constant, so that we may say we have

$$s = f(x + c_2,\ c,\ c_1) \;\ldots\ldots\ldots\ldots\ldots\ldots\ldots(13).$$

And in like manner we should have

$$s = F(y + c_3,\ c,\ c_1)\ldots\ldots\ldots\ldots\ldots\ldots\ldots(14),$$

where c_3 is another constant.

We have thus four constants c, c_1, c_2, c_3; these must be determined from the four conditions $x = 0$ when $s = 0$, $x = a$ when s has its given length, $y = h$ when $s = 0$, $y = k$ when s has its given length. The equation to the curve would be found by eliminating s between (13) and (14). Then from (11) we could find λ by ascribing any value we please to x; for instance, if we put $x = a$, we see that λ is the value of $c\dfrac{\sqrt{(1+p^2)}}{p}$ when $x = a$, and this of course is known since the equation to the curve is known and c is known.

258. The form in which we have left δu at the end of Art. 256 suggests to us to ask if we can apply Jacobi's method here. It is obvious that the occurrence of s in the problem of Art. 239 renders the problem different from those which merely involve y and its differential coefficients for which Jacobi's method is specially constructed.

By (13) we have

$$s = f(x + c_2, c, c_1).$$

Now let z stand for $\frac{ds}{dc_2}$ or for $\frac{ds}{dc_1}$; then according to the principles of Jacobi's method, we see that if z be put for δs in

$$\phi''(s)\,\delta s - \frac{d}{dx}\left(\frac{c}{p^3}\frac{d\delta s}{dx}\right) = 0 \ldots\ldots\ldots\ldots\ldots\ldots(15),$$

the equation is satisfied. At first sight we might also suppose that if the value $\frac{ds}{dc}$ were put for δs, the equation (15) would be satisfied; so that apparently *three* forms could be found for δs; but it is obvious that there cannot be more than *two* different forms, inasmuch as a linear differential equation of the second order can only have two particular forms of solution. And in fact, since the quantity c itself occurs in (15) it will be found on examination that we are not justified in concluding that $\frac{ds}{dc}$ will be a solution of (15).

It is easy to verify that $\frac{df}{dc_2}$ is a particular solution; for by (13) we have

$$\frac{df}{dc_2} = \frac{df}{dx} = \frac{ds}{dx}.$$

We have to shew then that (15) is satisfied when $\frac{ds}{dx}$ is put for δs.

Now $$\phi'(s) = \frac{c}{p^2\sqrt{(1+p^2)}}\frac{dp}{dx};$$

therefore $$\phi''(s)\frac{ds}{dx} = \frac{d}{dx}\left\{\frac{c}{p^2\sqrt{(1+p^2)}}\frac{dp}{dx}\right\},$$

that is $$\phi''(s)\frac{ds}{dx} = \frac{d}{dx}\left\{\frac{c}{p^3}\frac{d\frac{ds}{dx}}{dx}\right\} \qquad (16).$$

This gives the required verification.

Since then we have a particular solution of (15) we can easily find the general solution. Denote it by $t\frac{ds}{dx}$: substitute in (15), and use (16), thus we obtain

$$\frac{\frac{d^2t}{dx^2}}{\frac{dt}{dx}} = -\frac{2\frac{d^2s}{dx^2}}{\frac{ds}{dx}} - \frac{\frac{d}{dx}\left(\frac{c}{p^3}\right)}{\frac{c}{p^3}};$$

therefore $$\frac{dt}{dx}\left(\frac{ds}{dx}\right)^2\frac{c}{p^3} = \text{constant}.$$

Hence $t = b_1\int\frac{p^3\,dx}{1+p^2}$, where b_1 is a constant.

This shews that the general solution of (15) is

$$b_1\sqrt{(1+p^2)}\int\frac{p^3\,dx}{1+p^2} + b_2\sqrt{(1+p^2)},$$

where b_2 is another constant.

259. The result just obtained is exactly equivalent to what we should obtain by the method indicated at the beginning of the preceding Article. For denoting by β_1 and β_2 arbitrary constants we expect to find for the general solution

$$\beta_1\frac{ds}{dc_1} + \beta_2\frac{ds}{dc_2}.$$

Now $$x + c_2 = \int\frac{\{c_1 - \phi(s)\}\,ds}{[c^2 + \{c_1 - \phi(s)\}^2]^{\frac{1}{2}}};$$

hence
$$\frac{dx}{ds}\frac{ds}{dc_1} = \int \frac{c^2 ds}{[c^2 + \{c_1 - \phi(s)\}^2]^{\frac{3}{2}}},$$

and as $p = \dfrac{c}{c_1 - \phi(s)}$ we have

$$\frac{dx}{ds}\frac{ds}{dc_1} = \frac{1}{c}\int \frac{p^3 ds}{(1+p^2)^{\frac{3}{2}}} = \frac{1}{c}\int \frac{p^3 dx}{1+p^2};$$

therefore
$$\frac{ds}{dc_1} = \frac{1}{c}\frac{ds}{dx}\int \frac{p^3 dx}{1+p^2}.$$

And $\dfrac{dx}{ds}\dfrac{ds}{dc_2} + 1 = 0$, so that $\dfrac{ds}{dc_2} = -\dfrac{ds}{dx}$.

Hence $\beta_1 \dfrac{ds}{dc_1} + \beta_2 \dfrac{ds}{dc_2}$ is equivalent to

$$b_1 \frac{ds}{dx}\int \frac{p^3 dx}{1+p^2} + b_2 \frac{ds}{dx}.$$

We shall find also that

$$\frac{ds}{dc} = -\frac{1}{c}\frac{ds}{dx}\int \frac{p^2 dx}{1+p^2},$$

so that $\dfrac{ds}{dc}$ would differ in form from $\dfrac{ds}{dc_1}$ and $\dfrac{ds}{dc_2}$, and would not furnish a solution of (15).

CHAPTER XII.

MULTIPLE SOLUTIONS.

260. We have given numerous examples of discontinuous solutions in which the discontinuity could be traced to the circumstance that some condition or conditions had been imposed on the problem either explicitly or implicitly. Another kind of discontinuity has also presented itself, which may be considered to arise from the circumstance that the fundamental equation furnished by the Calculus of Variations has more than one solution. Examples may be seen in Chapters I, VI, IX, and XI. I will call these solutions *Multiple Solutions*; and will now proceed to some remarks on them.

261. Let v be any function of x and y and the differential coefficients of y with respect to x. Let $u = \int_{x_0}^{x_1} v\,dx$; and suppose we seek for a maximum or minimum of u. We obtain in the usual way to the first order

$$\delta u = L_1 - L_0 + \int_{x_0}^{x_1} M\,\delta y\,dx,$$

where M denotes a certain function of x and y, and the differential coefficients of y with respect to x; also L_1 and L_0 are the values when for x we put respectively x_1 and x_0 of a certain function L of x, y, and δy, and of the differential coefficients of y and δy with respect to x.

To make δu always $= 0$ we must have $M = 0$, and also $L_1 = 0$ and $L_2 = 0$.

Now suppose that M breaks up into factors, say two factors N and P; and let us enquire if we can use both factors in *one* solution of the same problem.

For example, suppose that from $x = x_0$ to $x = \xi$ we take $N = 0$; and then from $x = \xi$ to $x = x_1$ take $P = 0$.

Then $N = 0$ must be so taken as to make $L = 0$ when $x = x_0$, and $P = 0$ must be so taken as to make $L = 0$ when $x = x_1$. And there will now be a term in δu which we may denote by $L_2 - L_3$; where L_2 denotes what L becomes when we put ξ for x and use the relation $N = 0$, and L_3 denotes what L becomes when we put ξ for x and use the relation $P = 0$. In general, then, the coefficients of δp and the higher variations in L_2 and L_3 must separately vanish, for the δy will be the same in both, but the δp and the higher variations will not be the same. If all the conditions thus arising can be satisfied, the discontinuous solution formed from the combination of $N = 0$ and $P = 0$ is admissible; subject of course to the necessity of ascertaining that the term of the second order in δu is positive for a minimum and negative for a maximum.

262. Although the preceding Article shews very distinctly that the kind of discontinuity there contemplated may exist, yet good examples of it do not appear to present themselves readily. It is easy to construct examples in which the quantity v itself has a factor raised to the second power, or to a higher power, which will occur in M, and so by being equated to zero may furnish part of a solution. For instance, in Art. 111, we have y^2 occurring in v, and y in M; and $y = 0$ is used as part of the solution.

There is one remark of importance with respect to the theory of the preceding Article which ought to be made. The differential coefficient of the highest order which occurs in M presents itself in the *first* degree; thus of course it will not enter into both N and P. Hence one of the two differential equations $N = 0$,

$P=0$ is of a lower order than $M=0$; and therefore, by integrating it we shall not obtain so many arbitrary constants as we require to make the term in δu vanish which is outside the integral sign: thus we shall not succeed in obtaining the discontinuous solution except under specially favourable circumstances. It may be observed, that although we appear to save one equation of condition from the circumstance that δy has but one meaning at the point of discontinuity, yet we have on the other hand the condition to satisfy that $N=0$ and $P=0$ shall give the *same value* of y at this point. Similarly δp will have but one meaning at the point of discontinuity if we make the two curves touch there; [provided we assume that there is no break of direction at the point: see Art. 20.] And so on.

263. There are however cases of multiple solutions of a somewhat different kind, which are not devoid of interest; I mean such as those considered in Chapter I. Let us take another example resembling that of Art. 15.

Suppose that $\phi(q) = a^4 q^2 + \frac{b^4}{q^2}$; and let it be required to find a curve joining two fixed points for which $\int \phi(q)\, dx$ is a minimum.

Take the first fixed point for the origin; and let x_1 and y_1 be the co-ordinates of the other fixed point.

Let $u = \int \phi(q)\, dx$; the limits being fixed. Then to the first order we have

$$\delta u = \delta p\, \phi'(q) - \delta y \frac{d}{dx} \phi'(q) + \int \left\{\frac{d^2}{dx^2} \phi'(q)\right\} \delta y\, dx,$$

the whole taken between the fixed limits.

Thus the general solution of the problem is given by

$$\phi'(q) = C_1 x + C_2,$$

where C_1 and C_2 are constants.

Now δy is zero at the limits; but as δp is not, it will be necessary that $\phi'(q)$ should vanish at both limits: this leads to

$$C_1 = 0 \text{ and } C_2 = 0;$$

therefore $\phi'(q) = 0$; that is, in this case

$$a^4 q - \frac{b^4}{q^3} = 0.$$

Hence $q = \pm \frac{b}{a}$; and therefore we have

$$y = \frac{\beta x^2}{2} + \lambda x + \mu,$$

where $\beta = \pm \frac{b}{a}$, and λ and μ are constants.

Then, in order that the curve may pass through the given points, we must have $\mu = 0$, and λ must be found from

$$y_1 = \frac{\beta x_1^2}{2} + \lambda x_1.$$

The term of the second order in δu is

$$\frac{1}{2}\int \phi''(q)\,(\delta q)^2\,dx,$$

and this is necessarily positive; so that we have a minimum.

In fact we have here not only a *minimum*, but we have the *least* possible value of the proposed integral. For

$$a^4 q^2 + \frac{b^4}{q^2} = \left(a^2 q - \frac{b^2}{q}\right)^2 + 2a^2 b^2;$$

and this has its *least* possible value when $q = \pm \frac{b}{a}$; and so the integral has then its least possible value.

But it must be observed that we can give discontinuity to our solution; and make it consist of a broken or zigzag path. Instead of going along one parabola from the first fixed point to the second, we can take arcs of different parabolas, all being given by an equation of the form

$$y = \frac{\beta x^2}{2} + \lambda x + \mu,$$

where β may be at our pleasure either $\frac{b}{a}$ or $-\frac{b}{a}$, and λ and μ are constants. It is obvious in fact that we get the same value for the integral as before.

This amount of discontinuity might indeed have been anticipated; for since u involves only q^2, we might have expected that so long as the value of q was numerically unchanged the integral would remain unchanged.

264. Suppose now that we modify the problem by making the curve touch given straight lines at both the fixed points.

In this case δp is zero at the fixed points; thus we have no longer $C_1 = 0$ and $C_2 = 0$. Hence a solution is to be obtained from

$$\phi'(q) = C_1 x + C_2;$$

the two constants which are here expressed, and the two more which will enter in the value of y in terms of x, must then be determined by the conditions that the curve is to pass through two fixed points, and touch fixed straight lines at those points.

This solution will give us a *minimum*, but it will not give us the *least* value of the proposed integral. For we can still find a discontinuous solution; we may suppose it composed of the two parabolas

$$y = \frac{\beta}{2} x^2 + \gamma x,$$

and

$$y - y_1 = \frac{\beta}{2}(x^2 - x_1^2) + \lambda(x - x_1),$$

where $\beta = \pm \frac{b}{a}$, and we may use either value in either equation.

That is, we start from the origin on the first parabola, and continue on it up to the point where the two parabolas meet; and then we proceed along the second parabola to the second fixed point. The constant γ must be determined so as to make the first parabola touch the given straight line at the origin; and the constant λ must be determined so as to make the second parabola touch the other given straight line at the second fixed point. All the conditions for a minimum are satisfied by this solution; and in fact we see, as before, that it gives us the *least* possible value of the proposed integral.

265. It is obvious that a result of a similar kind to that in the preceding Article will hold whenever we seek the maximum or minimum value of an integral which involves nothing except one differential coefficient. For instance, let r stand for $\frac{d^3y}{dx^3}$; required a maximum or minimum of $\int \phi(r)\, dx$ between fixed limits. Let u denote the integral; then to the second order

$$\delta u = \int \left\{ \phi'(r)\, \delta r + \frac{\phi''(r)\, (\delta r)^2}{2} \right\} dx.$$

Now we do not assert that we *must* have $\phi'(r) = 0$; because it is not obviously certain that δr can have either sign consistently with such conditions as may be imposed at the limits; but we can always try if $\phi'(r) = 0$ will give a solution. If $\phi'(r)$ breaks up into factors, we can combine two factors; say $r = \alpha$ and $r = \beta$ are thus deduced. Then from each of these we can get a relation between x and y and *three* arbitrary constants; thus on the whole we have *six* arbitrary constants, which is the number we should get from the ordinary continuous solution, namely, from $\frac{d^3}{dx^3}\phi'(r) = 0$. Thus we can in general make the discontinuous solutions satisfy as many conditions as the ordinary continuous solution. Compare Art. 262. The solution gives a maximum if $\phi''(r)$ be negative, and a minimum if $\phi''(r)$ be positive.

266. There is another way in which it is conceivable that discontinuity might occur in problems of the Calculus of Varia-

tions. Take the general equation $M=0$ of Art. 261, and try if we can employ two solutions of it, one with one set of arbitrary constants, and the other with another set. Of course the equation $M=0$ of Art. 261 will be satisfied whatever be the values of the arbitrary constants. But when we consider the term L, we shall find that it will in general be impossible to satisfy all the conditions relating to this term with different sets of arbitrary constants.

267. The remark in the preceding Article, however, must not lead us to suppose that we can never make use of a combination of two solutions of the equation $M=0$, which differ in the constants involved: the next Chapter will furnish an illustration of the possibility of such a combination. In Chapter VII. we have also Examples of such combination; take, for instance, that in Art. 143. And a very instructive case, though of somewhat different kind, occurs in Art. 41: here, in fact, the equation $M=0$ is satisfied by $r=\frac{a}{\cos(\theta-\beta)}$, where β is a constant which is zero for one part of the solution, and $\frac{\pi}{2}$ for another part; but instead of shewing that the terms arising from L vanish, we shew that they are essentially negative. See also Art. 64.

CHAPTER XIII.

AREA BETWEEN A CURVE AND ITS EVOLUTE.

268. REQUIRED a curve connecting two fixed points such that the area between the curve, its evolute, and the radii of curvature at its extremities may be a minimum.

This is a well-known problem, which has however hitherto been very imperfectly discussed.

We will suppose the curve to be concave to the axis of x, so that q is negative.

Let $$u = \int \frac{(1+p^2)^2}{-q}\, dx,$$

the integral being supposed to be taken between fixed limits.

Then to the second order

$$\delta u = -\int \left\{ \frac{4p(1+p^2)}{q}\, \delta p - \frac{(1+p^2)^2}{q^2}\, \delta q \right\} dx$$

$$-\frac{1}{2} \int \left\{ \frac{4(1+3p^2)}{q} (\delta p)^2 - \frac{8p(1+p^2)}{q^2}\, \delta p\, \delta q + \frac{2(1+p^2)^2}{q^3} (\delta q)^2 \right\} dx.$$

The term of the first order in δu becomes by the usual transformation

$$\frac{(1+p^2)^2}{q^2}\, \delta p - \left\{ \frac{4p(1+p^2)}{q} + \frac{d}{dx} \frac{(1+p^2)^2}{q^2} \right\} \delta y$$

$$+ \int M \delta y\, dx,$$

where M stands for $4\frac{d}{dx}\frac{p(1+p^2)}{q}+\frac{d^2}{dx^2}\frac{(1+p^2)^2}{q^2}$; and the whole expression is to be taken between the limits.

We must then put $M=0$. As u involves only p and q we can by the ordinary method obtain the second integral of the differential equation $M=0$, namely,

$$-\frac{(1+p^2)^2}{q}=\frac{(1+p^2)^2}{q^2}q+C_1p+C_2,$$

where C_1 and C_2 are constants.

Thus
$$\frac{(C_1p+C_2)q}{(1+p^2)^2}=-2.$$

Let s denote the arc of the curve measured from a fixed point, ρ the radius of curvature; then the last equation gives

$$C_1\frac{dy}{ds}+C_2\frac{dx}{ds}=2\rho.$$

Let ϕ be the angle which the tangent to the curve makes with the axis of x; and assume $C_1=k\sin\beta$, $C_2=k\cos\beta$; thus

$$\rho=\frac{k}{2}\cos(\phi-\beta).$$

This is the well-known intrinsic equation to a cycloid; the radius of the generating circle being $\frac{k}{8}$, that is $\frac{\sqrt{(C_1^2+C_2^2)}}{8}$.

Since the extreme points are supposed fixed, the integrated part of the term of the first order in δu reduces to

$$\left\{\frac{(1+p^2)^2}{q^2}\delta p\right\}_1-\left\{\frac{(1+p^2)^2}{q^2}\delta p\right\}_0.$$

If the tangents at the fixed points have given directions, then $\delta p_1=0$, and $\delta p_0=0$; and thus the integrated part of δu vanishes.

If the tangents at the fixed points have not given directions we must have $\frac{(1+p^2)^2}{q^2}$ equal to zero at the fixed points; this requires that the radius of curvature should vanish at these points, and so the cycloid must have cusps at these points.

The term of the second order in δu may be put in the form

$$\int -\frac{1}{q}\left\{2(1+p^2)(\delta p)^2 + \left(2p\delta p - \frac{1+p^2}{q}\delta q\right)^2\right\} dx;$$

and since q is negative this term is essentially positive, and so we have a minimum.

269. Let us now consider the result more closely. Take, for example, the case in which the directions of the tangents at the fixed points are given. Suppose that an arc of a cycloid has been found which joins the two fixed points, and has its tangents at these points in the given directions, and has no cusp between the fixed points. Then this gives a *minimum* value of the area under consideration; that is, if we pass from the cycloid to any adjacent curve which can be reached by an admissible variation, the area is increased. But it does not follow that we have thus obtained the *least* area. In fact by employing a series of arcs of a cycloid, or even of arcs of a circle, we may make the area as small as we please.

270. Suppose however that we limit the problem thus: required a curve so that the area under consideration may be a minimum under the conditions that the directions of the tangents at the fixed points are given, that there is to be no abrupt change of direction, and that the curve is never to be convex to the straight line which joins the fixed points.

It seems obvious that under these conditions there must be some curve which gives the *least* value of the area in question; and that this least value must be greater than zero. We should therefore admit that the required curve can be nothing else than the single arc of a cycloid. This of course rests on the assumption that the Calculus of Variations furnishes no other solution besides that at which we have already arrived.

271. Let us proceed to examine whether there is any trace of another solution.

If we put $C_1 = 0$ and $C_2 = 0$ in Art. 268 we have $\rho = 0$; thus a circle of indefinitely small radius may be considered a kind of solution. This may be employed in a certain case.

Suppose that the given directions of the tangents at the fixed points A and B make *obtuse* angles with the straight line AB.

We may consider the solution to consist of arcs of circles of infinitesimal radius at A and B joined on to a cycloid at the cusps. Thus the amount of the area is the same as if the angles at A and B had been given to be right angles, or as if there had been no condition as to these angles.

272. Let us now examine if there is any *multiple* solution of the kind noticed in Chapter XII.

We have to the first order

$$\delta u = J\delta y + K\delta p + \int M\delta y dx,$$

where $$J = -\left\{\frac{4p(1+p^2)}{q} + \frac{d}{dx}\frac{(1+p^2)^2}{q^2}\right\},$$

$$K = \frac{(1+p^2)^2}{q^2},$$

and M has the value given in Art. 268.

Thus if the solution could be composed of two arcs of cycloids it would be necessary that at the common point p, q, and $\frac{dq}{dx}$ should have respectively the same value for each arc. It remains to ascertain whether this is possible.

Take two different cycloids; select any point on one of them; then find the point on the other where the radius of curvature is of the same value as at the selected point on the first cycloid. Let the two cycloids be placed so that they may have a common tangent at the point where the radius of curvature has the same value for both. Thus we secure that p and q have the same value for the two arcs at the common point; but we have still to determine whether $\frac{dq}{dx}$ can have also the same value for the two arcs which is necessary.

For one cycloid let a_1 be the radius of the generating circle, ρ_1 the radius of curvature, s_1 the arc measured from the vertex, ϕ_1 the angle between the direction of ρ_1 and that of the normal at the vertex; then $\rho_1 = 4a_1 \cos\phi_1$. With a similar notation for the other cycloid we have $\rho_2 = 4a_2 \cos\phi_2$. Now since there is a common value of p and a common value of q at the common point, there will also be a common value of $\frac{dq}{dx}$ provided $\frac{d\rho_1}{ds_1} = \frac{d\rho_2}{ds_2}$. This

leads to $$4a_1 \sin\phi_1 \frac{d\phi_1}{ds_1} = 4a_2 \sin\phi_2 \frac{d\phi_2}{ds_2},$$

that is to $$\frac{4a_1 \sin\phi_1}{\rho_1} = \frac{4a_2 \sin\phi_2}{\rho_2};$$

and since by supposition $\rho_1 = \rho_2$ we have $4a_1 \sin\phi_1 = 4a_2 \sin\phi_2$ if ρ_1 and ρ_2 are different from zero. Thus if ρ_1 and ρ_2 are different from zero, we must have $a_1 \sin\phi_1 = a_2 \sin\phi_2$ as well as

$$a_1 \cos\phi_1 = a_2 \cos\phi_2;$$

so that we get $a_1 = a_2$ and $\phi_1 = \phi_2$; thus our two cycloids really *coincide*.

But if we suppose $\phi_1 = \frac{\pi}{2}$ and $\phi_2 = \frac{\pi}{2}$ we have ρ_1 and ρ_2 each zero; and two different cycloids meeting at a cusp, and having a common tangent there present themselves.

Thus let AC be an arc of one cycloid, and BC of another, C being a cusp in each, and the arcs having a common tangent at C; then the conditions of the problem are satisfied by ACB, provided the tangents at A and B have the right directions.

It is not necessary that a_1 should be equal to a_2: so that AC and CB may not only be different cycloids, but generated by different circles.

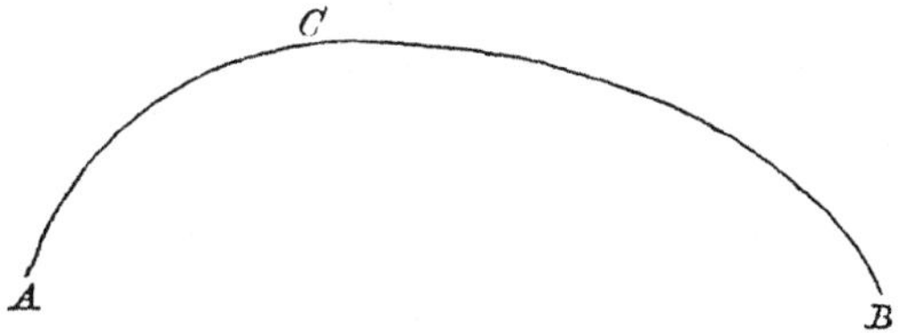

This result is of course subject to any suspicion which may arise from the fact that q is infinite at C; but as q only occurs in the denominators of our expressions the difficulty does not seem serious. Indeed this is not peculiar to our discontinuous solution, but occurs in the ordinary solution, as in Art. 268, inasmuch as q is infinite at the cusp at which we there arrived.

273. As a simple application of the preceding investigations suppose that the tangents at A and B are to be at right angles to AB. Then one solution is supplied by a cycloid having cusps at A and B, as in Art. 268. Another solution is supplied according to Art. 272 by two half cycloids having vertices at A and B and joined at a cusp at C, where they must have a common tangent: the two half cycloids in this case are generated by the same circle. Each solution seems to give a minimum; the area in the former is however less than the area in the latter: though it may indeed be said that part of the area in the latter case is reckoned twice.

274. Let us now proceed to impose conditions. For example, suppose that a certain given point is not to fall outside the curve. Of course it may happen that the solution already obtained is still applicable; this will be the case if the given point is not too far from AB. Or it may happen that of the two solutions of Art. 273 the latter is applicable but not the former. We should in most cases however have to seek another solution. The required curve will then pass through the given point; and we must be prepared for discontinuity there.

Corresponding to the term $K\delta p$ we now get $(K\delta p)_2 - (K\delta p)_3$, where the subscript 2 refers to the end of one arc and the subscript 3 to the beginning of the other arc, at the common point. We shall not ensure that this vanishes unless we have $\delta p_2 = \delta p_3$, and $K_2 = K_3$. [The former condition implies that we have imposed the restriction that there is to be no abrupt change of direction.] Thus the two arcs must touch at the common point. Then the latter condition requires that the radius of curvature should be the same for the two arcs at the common point.

The integrated part of the variation thus reduces to

$$(J\delta y)_2 - (J\delta y)_3.$$

Since $\delta y_2 = \delta y_3$, $p_2 = p_3$, and $q_2 = q_3$, this becomes

$$\left\{\frac{2(1+p^2)^2}{q^3}\delta y\right\}\left\{\left(\frac{dq}{dx}\right)_2 - \left(\frac{dq}{dx}\right)_3\right\}.$$

The first factor may have indifferently 2 or 3 for subscript. Of course if we could have $\left(\frac{dq}{dx}\right)_2 = \left(\frac{dq}{dx}\right)_3$ this term would vanish; but by supposition we cannot in general satisfy this additional condition. Since δy is necessarily positive at the point we only require that $\left(\frac{dq}{dx}\right)_2 - \left(\frac{dq}{dx}\right)_3$ shall be positive; and then the term will be positive, so that we shall have a minimum. Thus there will in general be discontinuity at the point which is not to be excluded by the solution; we shall have two arcs of cycloids which meet at the point, have a common tangent there, and a common value of the radius of curvature.

275. Let us now add the condition to the problem that the required curve is never to go beyond the boundary defined by some given curve.

We are certain that the required solution can be composed of nothing but an arc or arcs of a cycloid, and a part or parts of the given curve.

Of course if an arc of a cycloid can be obtained which falls entirely within the given boundary that arc is the solution. Suppose however that this is not the case. Let the value of the expression denoted by M in Art. 268 be found for the given curve.

If through any part of the given curve M is negative that part is available for our purpose; and the solution may consist of that part combined with an arc or arcs of a cycloid. For the value of $\int M\delta y\,dx$ will be zero for the cycloidal part, and it will be positive for the part of the given curve, unless indeed δy be zero throughout this part.

At the point or points of junction of the given curve with the cycloids the values of p and q must be the same for the given curve and for the cycloid, in order that the terms corresponding to $K\delta p$ may vanish, as in Art. 274: and from considering the terms corresponding to $J\delta y$ we draw an inference in the manner of that Article.

276. It is obvious however that even if M is negative for part of the given curve it may be impossible to satisfy all the other conditions which, as we have just seen, must hold in order that this part may contribute to the required solution. If M is positive throughout the given curve no part of that curve is available towards the solution. Thus we have still to find the solution applicable to such cases.

The subject will be sufficiently illustrated by taking a simple instance. Suppose then that the required curve is never to pass beyond a straight line; if this straight line is sufficiently near to the straight line AB, the solutions already given become inapplicable.

The following appears to be the solution. If the directions of the tangents at A and B are not given we must have cusps

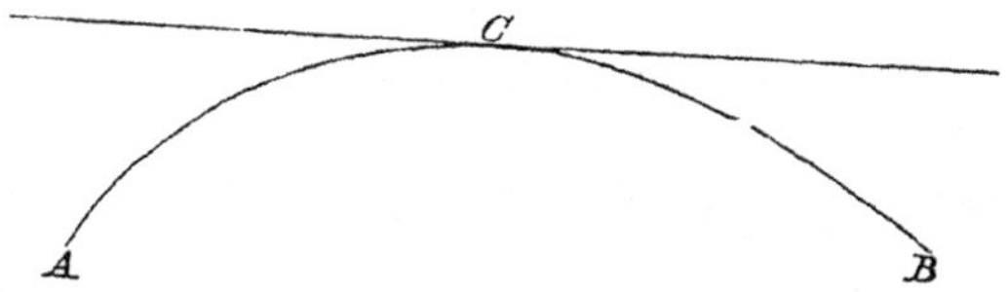

there; AC and BC are arcs of cycloids, which meet and have the given straight line for a common tangent at some point C. The arcs must have a common value for the radius of curvature at *C*. Also a condition must be satisfied relative to the term $J\delta y$, of the kind considered in Art. 274.

If the directions of the tangents at *A* and *B* are given, we may still employ this solution provided the tangents are to make with *AB* angles which are not less than those obtained in this solution; for we may suppose that arcs of a circle of infinitesimal radius are supplied at *A* and *B*. But if the solution cannot be employed we must abandon the condition of having cusps at *A* and *B*, and use instead that of the given directions for the tangents of the arcs.

CHAPTER XIV.

MISCELLANEOUS OBSERVATIONS.

277. I PROPOSE in this Chapter to discuss some topics which are connected more or less closely with our general subject, or with the problems which we have given to illustrate it.

278. The problem which I have discussed in Chapter V. is also considered in the work on the Calculus of Variations by Moigno and Lindelöf; see their page 224: these writers advert to the discontinuous solutions which had been proposed, and pronounce them not completely satisfactory. The substance of their remarks is the following: the integral $\int y\sqrt{(1+p^2)}\,dx$ will be in part negative, if we allow y to be negative; and the negative part and the positive part may each become as large as we please, and hence we ought not to be surprised that $\int y^2 dx$ has no maximum. Even if we admit that the remarks of Moigno and Lindelöf furnish some explanation of the apparent failure of the ordinary theory in this example, yet they do not in any degree discourage us from attempting to solve the problem with the condition annexed that *y shall be always positive:* indeed it might be said that this condition would naturally be understood if not explicitly stated. But even if the condition that y shall be always positive, instead of being naturally suggested, were arbitrarily imposed, there would be sufficient interest in the problem to justify the discussion of it.

The history of mathematics shews that much may be gained by "striving against self-imposed difficulties;" see De Morgan's preface to Ramchundra's Treatise on Problems of Maxima and Minima: and I hope that the present researches will furnish at least some hints for the extension and improvement of the Calculus of Variations.

279. There is one point connected with the term of the second order in a variation which might naturally cause a difficulty, and which should therefore be noticed. It will be found from examples that the term of the second order in a variation may appear in a more or less simple form according as we take one or the other of the variables for the independent variable. It might at first be supposed that the two forms ought to be absolutely equivalent; such however is not the case. To take a simple example; let

$$u = \int y^2 dx,$$

change y into $y + \delta y$, then

$$\delta u = 2\int y\,\delta y\,dx + \int (\delta y)^2\,dx.$$

Now take y for the independent variable instead of x, and so put u in the form $\int y^2 \varpi dy$, where ϖ stands for $\frac{dx}{dy}$. Change x into $x + \delta x$, then

$$\delta u = \int y^2 \delta\varpi dy$$

$$= y^2\delta x - 2\int y\delta x dy.$$

Thus while in the former value of δu there is a term of the second order, as well as a term of the first order, in the latter value of δu there is no term of the second order.

280. In general let v denote any function of x, y, and the differential coefficients of y with respect to x. Let $u = \int v dx$. Change y into $y + \delta y$, and let the part of δu which is of the first

order, and which remains under the integral sign after the usual transformation, be denoted by $\int M\,\delta y\,dx$. Now take y for the independent variable instead of x. Change x into $x + \delta x$, and let the part of δu which is of the first order, and which remains under the integral sign after the usual transformation, be denoted by $\int N\,\delta x\,dy$. Then it is known that $N = -M$; this result was obtained at a very early date in the history of the subject: see Todhunter's *History of the Calculus of Variations*, page 64.

Still we cannot say that the term of the first order in the one case is equal to the term of the first order in the other case; the terms differ in general by a term of the second order.

A geometrical illustration will make the matter clear. Suppose that the variation δy is always assigned by the relation

$$f(y + \delta y, x) = 0 \quad\text{.......................}\quad (1),$$

so that $y + \delta y$ is always the ordinate of a certain curve corresponding to the abscissa x. Then if we wish the same curve to be obtained by variation in the second case, as in the first, the variation δx must be assigned by the relation

$$f(y, x + \delta x) = 0 \quad\text{.......................}\quad (2).$$

From (1) and (2) we deduce

$$\delta y = -p\delta x + \omega \quad\text{.......................}\quad (3),$$

where ω is of the order of squares and products of δy and δx.

Then from (3) it follows, as we have stated, that $\int N\delta x dy$ and $\int M\,\delta y\,dx$ will in general differ by a quantity of the second order.

In like manner the term of the second order in one form of δu will differ from the term of the second order in the other form of δu, the difference may be of the second order, as we see by the simple example of Art. 279; and in fact the difference will be in general of the second order because $\int N\,\delta x\,dy$ and $\int M\,\delta y\,dx$ differ by a term of the second order.

281. The parts of the terms of the first order which occur free from the sign of integration in the two forms of δu are connected, though not in so simple a manner, as the parts which remain under the integral sign. Suppose that v is a function of x, y, p, q. Let Y stand for $\frac{dv}{dy}$, Y_1 for $\frac{dv}{dp}$, and Y_2 for $\frac{dv}{dq}$; and let accents above the letters denote complete differential coefficients with respect to x. Then we know that if we change y into $y+\delta y$, we have to the first order

$$\delta u = \int (Y - Y_1' + Y_2'')\,\delta y\,dx$$

$$+ (Y_1 - Y_2')\,\delta y + Y_2\delta p\,;$$

both parts being taken between limits. Now, as we have seen,

$$\delta y = -p\delta x + \omega,$$

where ω is of the second order; so that to the first order we have

$$\delta y = -p\delta x, \text{ and } \delta p = -q\delta x - p^2\delta\varpi,$$

where $\delta\varpi$ is equivalent to $\frac{d\,\delta x}{dy}$. Substituting we have to the first order

$$\delta u = -\int (Y - Y_1' + Y_2'')\,p\delta x\,dx$$

$$-\{p\,(Y_1 - Y_2') + q\,Y_2\}\,\delta x - Y_2 p^2\delta\varpi.$$

Thus if v were expressed in terms of y, x, and the differential coefficients of x with respect to y, and then x changed into $x+\delta x$, the coefficients of δx and $\delta\varpi$ in the part of δu which is free from the integral sign will be equal to the corresponding coefficients just expressed.

For an example of the different forms in which the term of the second order in a variation may appear, we can take the brachistochrone problem. Put $u = \int \frac{\sqrt{(1+p^2)}.dx}{\sqrt{y}}$; then the term of the second order in δu takes the form

$$\frac{1}{2}\int\left\{\frac{3(1+p^2)^{\frac{1}{2}}}{4y^{\frac{5}{2}}}(\delta y)^2 - \frac{p\delta p\delta y}{y^{\frac{3}{2}}(1+p^2)^{\frac{1}{2}}} + \frac{(\delta p)^2}{y^{\frac{1}{2}}(1+p^2)^{\frac{3}{2}}}\right\}dx.$$

Now take y for the independent variable instead of x; then u becomes $\int \frac{\sqrt{(1+\varpi^2)}\,dy}{\sqrt{y}}$; and the term of the second order in δu is

$$\frac{1}{2}\int\frac{(\delta\varpi)^2\,dy}{y^{\frac{1}{2}}(1+\varpi^2)^{\frac{3}{2}}}.$$

282. Suppose we wish to find a maximum or minimum of $\int \chi(y)\,\phi(p)\,dx$; this is substantially the same problem as that in Art. 163; but it may be useful to give the necessary formulæ explicitly.

Denote the integral by u; then to the second order

$$\delta u = \int\left\{\chi'(y)\,\phi(p)\,\delta y + \chi(y)\,\phi'(p)\,\delta p\right\}dx$$

$$+\frac{1}{2}\int\left\{(\delta y)^2\chi''(y)\,\phi(p) + 2\delta y\,\delta p\,\chi'(y)\,\phi'(p) + (\delta p)^2\chi(y)\,\phi''(p)\right\}dx.$$

The term of the first order is to be transformed in the usual way, and the part under the integral sign made to vanish: thus we get

$$\chi(y)\,\{\phi(p) - p\phi'(p)\} = \text{a constant} \ldots\ldots\ldots\ldots\ldots(1).$$

Also we have

$$\int \delta y\,\delta p\,\chi'(y)\,\phi'(p)\,dx = \frac{1}{2}(\delta y)^2\chi'(y)\,\phi'(p)$$

$$-\frac{1}{2}\int(\delta y)^2\,\{\chi''(y)\,p\phi'(p) + \chi'(y)\,\phi''(p)\,q\}\,dx,$$

so that the term of the second order in δu becomes

$$\frac{1}{2}(\delta y)^2 \chi'(y)\,\phi'(p) + \frac{1}{2}\int (\delta p)^2 \chi(y)\,\phi''(p)\,dx$$

$$+\frac{1}{2}\int (\delta y)^2 \{\chi''(y)\,[\phi(p) - p\phi'(p)] - \chi'(y)\,\phi''(p)\,q\}\,dx.$$

Sometimes we may be able to determine the sign of the term of the second order in δu from the expression just obtained.

If the constant in (1) is zero we must have p constant, whether we take $\chi(y) = 0$, or $\phi(p) - p\phi'(p) = 0$; so that $q = 0$, and this effects a great simplification in the term of the second order. If we take $\phi(p) - p\phi'(p) = 0$ the whole of the second line in the last expression for the term of the second order in δu vanishes.

If equation (1) is to hold for any value of y which makes $\chi(y)$ vanish, the constant must in general be zero.

Let us now consider the application of Jacobi's method to the term of the second order in δu in this case. Denote the constant in (1) by c_1. From this equation y is some function of p and c_1, say

$$y = f(p, c_1) \qquad (2),$$

and
$$x = \int \frac{dy}{p} + c_2 = \int \frac{1}{p}\frac{df(p, c_1)}{dp}\,dp,$$

say
$$x = F(p, c_1) + c_2 \qquad (3).$$

From (2) and (3) by eliminating p we obtain an equation between x, y, c_1 and c_2. We require to know the values of $\frac{dy}{dc_1}$ and $\frac{dy}{dc_2}$.

From (2) and (3)

$$\frac{dy}{dc_1} = \frac{df}{dc_1} + \frac{df}{dp}\frac{dp}{dc_1},$$

$$0 = \frac{dF}{dc_1} + \frac{dF}{dp}\frac{dp}{dc_1};$$

but $$p = \frac{df}{dp} q, \text{ and } 1 = \frac{dF}{dp} q;$$

therefore $$\frac{dy}{dc_1} = \frac{df}{dc_1} - p \frac{dF}{dc_1}.$$

Again from (2) and (3)

$$\frac{dy}{dc_2} = \frac{df}{dp} \frac{dp}{dc_2},$$

$$0 = \frac{dF}{dp} \frac{dp}{dc_2} + 1;$$

therefore $$\frac{dy}{dc_2} = -p.$$

Now $\frac{df}{dc_1}$ is to be obtained from (2) supposing p constant; or we may if we please obtain it from (1) which is equivalent to (2).

Thus $$\frac{\chi'(y)}{\chi(y)} \frac{df}{dc_1} = \frac{1}{c_1},$$

and therefore $$\frac{df}{dc_1} = \frac{\chi(y)}{c_1 \chi'(y)}.$$

Thus the quantity which was denoted by z in the account of Jacobi's method in Art. 24 becomes

$$B_1 \left\{ \frac{\chi(y)}{c_1 \chi'(y)} - p \frac{dF}{dc_1} \right\} - B_2 p,$$

where B_1 and B_2 are constants.

This is as far as we can carry the general process, because we cannot express $\frac{dF}{dc_1}$ while $\chi(y)$ remains quite general.

As an example suppose $\chi(y) = y^n$;

therefore, by (1), $$y = \left\{ \frac{c_1}{\phi(p) - p\phi'(p)} \right\}^{\frac{1}{n}}.$$

Thus we get $x = c_1^{\frac{1}{n}} \theta(p) + c_2$, where $\theta(p)$ denotes some function of p which does not contain c_1 explicitly. Here then

$$\frac{dF}{dc_1} = \frac{x - c_2}{nc_1},$$

and z becomes $$\frac{B_1}{nc_1}\{y - p(x - c_2)\} - B_2 p.$$

Hence the same interpretation with respect to the tangents at the extreme points of the curve which we may suppose to be required holds as in the case of $n = 1$: see Art. 29. And the reason is obvious; for by changing the constant we have in this case $y = c\psi(p)$, where c is a constant, and $\psi(p)$ is some function of p.

283. An important remark must be made with respect to *relative* maxima and minima which, so far as I know, is not to be found in treatises on the subject.

Suppose we require that $\int u dx$ shall be a maximum while $\int v dx$ is constant: then according to the usual theory we seek the maximum of $\int (u + av)\, dx$ where a is a constant.

Now when we come to the term of the second order in $\delta \int (u + av)\, dx$, it is quite conceivable that we may find it is not certainly negative; and thus we infer correctly that $\int (u + av)\, dx$ is not a maximum. But still it may be quite possible that $\int u dx$ may be a maximum: for the variations are really limited by the condition that $\int v dx$ is constant, and to this condition we pay no regard when we merely consider the term of the second order in $\int (u + av)\, dx$.

This remark is necessary in order to anticipate an objection which might be brought against some of our results.

For instance in Art. 96 we proposed a solution for a certain problem.

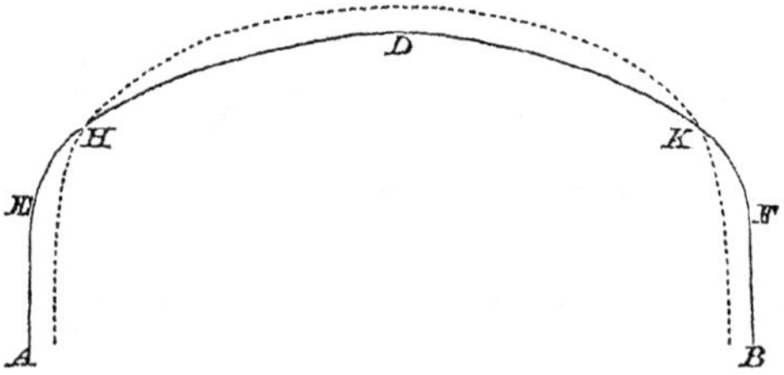

Let $AEDFB$ be the boundary of the solution there proposed. Suppose that the dotted boundary is obtained by giving an infinitesimal increment to the constant e of the investigation. Then there would not be a minimum of $\int \left(2y\sqrt{1+p^2} - \frac{y^2}{a}\right) dx$, because it is possible to draw between H and K a curve infinitesimally close to HDK satisfying the same differential equation. See the reasoning in Case II. of Art. 24.

Nevertheless this does not shew that the surface is not a minimum for a *given* volume: it is obvious that the curve here obtained by variation does not satisfy the condition of generating the *same* volume as the original curve. The volume is in fact increased by taking the dotted curve between H and K instead of the other curve.

I use the preceding only for illustration, so that it is not absolutely necessary for me to shew that such a line as the dotted line can be drawn. Nevertheless I think the following considerations will shew that such a line can be drawn. In the diagram of Art. 96 suppose a to remain constant, and e to receive a small increment. Then AE and DC increase, and AB decreases. Also the increase of AE bears a *finite* ratio to the decrease of AB; this shews that the varied place of E is *within* $AEDFB$, as I have drawn it.

Then the dotted line having crossed the other keeps above it up to D; this we see by the argument of Art. 97. Then by symmetry the dotted line crosses the other again at a point K which corresponds to H.

284. The Calculus of Variations is much stronger in its negative results than in its positive results, that is to say, we learn from it rather when a given expression has not a maximum or a minimum value than when it has.

Suppose that we have to find the maximum or minimum of $\int v dx$; denote this by u: then we obtain in the usual way

$$\delta u = L + \int M \delta y \, dx;$$

then we say that there can be no maximum or minimum, if δy be unrestricted, except when $M=0$; for if M is not $=0$ we can make δu positive or negative at our pleasure by taking δy suitably. But of course this does not ensure a maximum or a minimum without examining the term of the second order in δu.

There is, strictly speaking, only one case in which a perfectly definite result can be obtained, namely, when we know beforehand that there must be a maximum or that there must be a minimum. Take the case of the brachistochrone between fixed points; then the argument is as follows: we feel certain that there must be a line or lines of descent such that no other line of descent can be fallen through in *less* time; but if M is not zero the time of descent can be made less; therefore no curve can be the brachistochrone except a curve which satisfies the equation $M=0$ and passes through the fixed points; thus we feel certain that a curve must exist satisfying these conditions, and that it is the curve we require.

When this argument is put briefly it is often put inaccurately thus: we are sure that there can be no maximum in this case, and therefore the result must give a minimum. This is inaccurate, because we are not sure beforehand that there is no maximum in the technical sense of the word maximum; we see that the time of descent can be made as great as we please by suitably adjusting the line of descent, but this does not justify us in asserting that there can be no maximum.

But further; suppose that we examine the term of the second order in a variation. If we find, for instance, that this term is essentially positive, we can safely affirm that the relation which makes the term of the first order in the variation vanish does not give a maximum. But if we assert that the relation does give a minimum, we must bear in mind that this means a minimum with respect to *admissible variations.* Take for example the brachistochrone between fixed points. Suppose we draw close to the cycloid, which we obtain by making the term of the first order in the variation vanish, a line in the form of a series of indefinitely small steps, as in the diagram of Art. 205. Then the fact that the term of the second order in the variation is positive does not shew that the time down the cycloid is less than the time down the discontinuous figure, for our investigation is not applicable to such a variation as would be required in passing from the cycloid to the discontinuous figure: in such a passage δp would not be always indefinitely small. Of course it might be possible to give some special investigation for such a case, but certainly the case is not included in the ordinary methods of the Calculus of Variations.

When we assert then that a certain cycloid is the curve of quickest descent between two given points, the statement depends mainly on the fact that we feel certain beforehand that there is some curve which has the required property.

Similar remarks apply to other problems; so that we cannot by the aid of the Calculus of Variations assert that we have the least or the greatest value of a proposed integral unless we are certain beforehand that such a least or such a greatest value necessarily exists.

285. There is still another consideration. Suppose that we are examining a certain curve to see if it possesses a prescribed maximum property. It may happen that at a certain point of the curve p is infinite; the obstacle that thus arises in the use of the ordinary formulæ of the Calculus of Variations may prevent us from drawing the positive conclusion that there is a maximum: but such an obstacle may not prevent us from safely affirming that there is not a maximum. For we may of course apply the ordinary formulæ to such parts of the curve as have p finite; and

if the fundamental equation $M=0$ does not hold throughout such parts we are sure there is not a maximum. But if this equation does hold for the whole curve, and if the term of the second order in the variation is essentially negative, we cannot rely on our investigation so far as to assert that there is a maximum on account of the occurrence of the infinite value of p.

Similar remarks apply if any of the other quantities which occur become negative.

Take for example the brachistochrone between fixed points. If we make x the independent variable, x being measured horizontally, we have for the term of the second order in the variation the value given in Art. 281. But this is not trustworthy, for we have to vary $\frac{1}{y}$ and p which are both infinite at the starting point. If we make y the independent variable we have for the term of the second order the other value given in Art. 281; and this may be accepted without hesitation so long as ϖ is not infinite, that is, so long as we do not have to pass through the vertex of the cycloid in order to reach the second given point.

286. I have often spoken of the results which have been obtained as maxima or minima with respect to *admissible* variations. I will give another problem to illustrate this point.

A particle is to descend from one fixed point to another in a vertical plane, constrained by a smooth curve which is convex downwards: required the curve so that the integral $\int Pdt$ taken during the time of motion may be a minimum, where P denotes the pressure on the curve at the time t. Thus we may say that we require the whole pressure to be a minimum.

Take the highest point as the origin, and the axis of x vertically downwards; let v denote the velocity, ρ the radius of curvature, at the point (x, y); and let s denote the arc described up to this point.

Then $$P=\frac{v^2}{\rho}+g\frac{dy}{ds}, \text{ and } v=\sqrt{2gx}.$$

Thus the integral $=\int\left(\frac{v^2}{\rho}+g\frac{dy}{ds}\right)dt$

$$=\int\left(\frac{vds}{\rho}+\frac{gdy}{v}\right)=\int\left(\frac{\sqrt{2gx}\,q}{1+p^2}+\frac{gp}{\sqrt{2gx}}\right)dx.$$

Put u for $\int\left(\frac{2q\sqrt{x}}{1+p^2}+\frac{p}{\sqrt{x}}\right)dx$; then we require the minimum of u.

By the usual theory we must have

$$-\frac{d}{dx}\left\{\frac{1}{\sqrt{x}}-\frac{4pq\sqrt{x}}{(1+p^2)^2}\right\}+\frac{d^2}{dx^2}\,\frac{2\sqrt{x}}{1+p^2}=0\,;$$

therefore $$\frac{d}{dx}\,\frac{2\sqrt{x}}{1+p^2}+\frac{4pq\sqrt{x}}{(1+p^2)^2}-\frac{1}{\sqrt{x}}=C,$$

where C is a constant; thus

$$\frac{1}{(1+p^2)\sqrt{x}}-\frac{1}{\sqrt{x}}=C,$$

that is $$C=-\frac{p^2}{(1+p^2)\sqrt{x}}\;\ldots\ldots\ldots\ldots\ldots\ldots\ldots(1).$$

The term of the first order in δu which is free from the integral sign reduces to $\frac{2\sqrt{x}\,\delta p}{1+p^2}$, since the limits are fixed. This vanishes at the origin; to make it vanish at the second point we must have p infinite there, that is, the tangent must be horizontal. And if a be the value of x at the lowest point, we get $C=-\frac{1}{\sqrt{a}}$ from (1). Thus

$$\left(\frac{dy}{dx}\right)^2=\frac{\sqrt{x}}{\sqrt{a}-\sqrt{x}}\;\ldots\ldots\ldots\ldots\ldots\ldots\ldots(2)\,;$$

from this y must be found in terms of x.

Now let us examine the term of the second order in δu; this term is

$$\int\left\{\frac{2\,(3p^2-1)\,q\sqrt{x}\,(\delta p)^2}{(1+p^2)^3}-\frac{4p\sqrt{x}\,\delta p\delta q}{(1+p^2)^2}\right\}dx.$$

And $$\int \frac{p\sqrt{x}\,\delta p\delta q}{(1+p^2)^2}\,dx = \frac{p\sqrt{x}\,(\delta p)^2}{(1+p^2)^2} - \int \delta p\,\frac{d}{dx}\,\frac{p\sqrt{x}\,\delta p}{(1+p^2)^2}\,dx\,;$$

therefore $$\int \frac{p\sqrt{x}\,\delta p\delta q}{(1+p^2)^2}\,dx = \frac{p\sqrt{x}\,(\delta p)^2}{2\,(1+p^2)^2}$$

$$+\frac{1}{2}\int\left\{\frac{(3p^2-1)\,q\sqrt{x}}{(1+p^2)^3} - \frac{p}{2\,(1+p^2)^2\sqrt{x}}\right\}(\delta p)^2\,dx\,;$$

the part outside the integral sign vanishes at the limits. Thus the term of the second order in δu reduces to

$$\int \frac{p\,(\delta p)^2}{(1+p^2)^2\sqrt{x}}\,dx.$$

This is essentially positive; so that we may conclude that we have a minimum, subject of course to any doubt which may be produced by the occurrence of an infinite value of p. We may however safely assert that we have a minimum with respect to all admissible variations, with the condition that the tangent is to be always horizontal at the second point; for this condition makes $\delta p = 0$ when p is infinite, and so removes the difficulty which would otherwise exist.

But though the result obtained may be a minimum it does not give the *least* value of the proposed integral. For it is easy to see that if we pass from the curve determined by (2) to a step-shaped figure like that in Art. 205, the value of the integral is diminished. For by this transition we leave the element $g\frac{dy}{v}$ unchanged, and we change the element $\frac{v\,ds}{\rho}$ to zero: and thus we diminish the integral.

In fact the least value will be given by a discontinuous solution. The equation (1) admits of $p=0$ and $p=\infty$ as particular solutions. And the least value of the integral will be obtained by taking the portion of the axis of x from $x=0$ to $x=a$, and then a portion of the straight line $x=a$ up to the lower given point. That is, the required line is made up of two straight lines at right angles to each other.

That this discontinuous solution is really a minimum may be shewn thus: There must be some line which gives the least value to the whole pressure; this line must be a solution of (1); and the only solutions of (1) are the general solution given by (2), and the particular solutions $p=0$ and $p=\infty$. It is obvious that the general solution (2) produces a curve which is always convex downwards; and the discontinuous solution gives a less result than any curve which is convex downwards for two reasons: first, the element $\frac{vds}{\rho}$ is zero; and secondly, the element $\frac{gdy}{v}$, that is, $\frac{gdy}{\sqrt{(2ax)}}$ is replaced by the smaller element $\frac{gdy}{\sqrt{(2aa)}}$.

The condition that the curve is always to be convex downwards has been adopted to render the problem simple and definite. If this condition be not attached we may have P changing sign in the course of the integration. Moreover P would vanish throughout any arc of the parabola which the particle might freely describe under the action of gravity.

287. It may be useful to notice the formulæ which we obtain when we generalise the preceding problem.

Let $$u=\int\{q\phi(p)\,\psi(x)+p\chi(x)\}\,dx.$$

Then for a maximum or minimum value of u we must have

$$\phi(p)\,\psi'(x)-\chi(x)=\text{a constant}.$$

The term of the second order in δu reduces to

$$\frac{1}{2}\phi'(p)\,\psi(x)\,(\delta p)^2-\frac{1}{2}\int\phi'(p)\,\psi'(x)\,(\delta p)^2\,dx,$$

the whole taken between the limits.

288. In the problem discussed in Chapter V. I arrived at the result that no valid objection could be brought from Jacobi's method against the conclusion that a minimum value of the surface had been obtained. I shall now make some more remarks respecting the application of Jacobi's method to such problems.

When the value of y is fixed at the limits, we know by the investigation of Art. 23 that the term of the second order in the variation can be put in the form

$$\frac{1}{2}\int_{x_0}^{x_1}\left\{P\delta y-\frac{d}{dx}(Q\delta p)\right\}\delta y\,dx,$$

where x_0 and x_1 denote the limiting values of x.

Separate the integral into two parts, one extending from x_0 to ξ, and the other from ξ to x_1. Consider the former part; put $\delta y = tz$, where z is such that

$$Pz-\frac{d}{dx}\left(Q\frac{dz}{dx}\right)=0.$$

In the second part of the integral put $\delta y=\tau\zeta$, where ζ also satisfies the same differential equation as z does. The two solutions z and ζ are of course not necessarily the same; they may differ by taking different values of the arbitrary constants which will occur.

By thus separating the integral into two parts we shall find that the above term of the second order in the variation becomes

$$\frac{1}{2}\left\{-tQz^2\frac{dt}{dx}+\tau Q\zeta^2\frac{d\tau}{dx}\right\}_{x=\xi}$$

$$+\frac{1}{2}\int_{x_0}^{\xi}Q\left(\delta p-\frac{z'}{z}\delta y\right)^2dx+\frac{1}{2}\int_{\xi}^{x_1}Q\left(\delta p-\frac{\zeta'}{\zeta}\delta y\right)^2dx.$$

The terms under the integral signs are positive if Q is positive. The term outside the integral sign becomes by substitution

$$\frac{1}{2}Q\left\{-z\delta y\frac{d}{dx}\left(\frac{\delta y}{z}\right)+\zeta\delta y\frac{d}{dx}\left(\frac{\delta y}{\zeta}\right)\right\},$$

that is,

$$\frac{1}{2}Q\left(\frac{z'}{z}-\frac{\zeta'}{\zeta}\right)(\delta y)^2;$$

this assumes that there is no discontinuity at the point for which $x=\xi$, so that δp may have only one value at that point.

If then we can secure that $Q\left(\frac{z'}{z}-\frac{\zeta'}{\zeta}\right)$ is positive at this point, we are sure that the whole term of the second order in the variation will be positive.

Similarly we might proceed, if we wished to separate the integral into more than two parts. Now let us apply this process to the problem of Chapter V.

We shall separate the integral which represents the term of the second order in the variation into *three* parts. From the

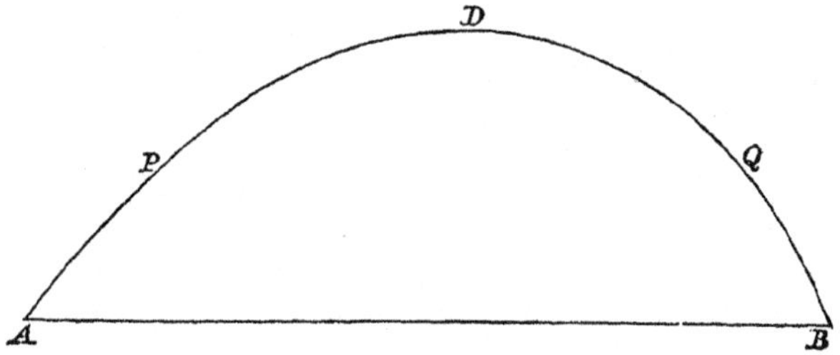

point A to some point between P and D we shall use the transformation $\delta y = tz$; and for z we shall take $C_1 p$, where C_1 is a constant. Then from the point thus reached to a second point between D and Q we shall use the transformation $\delta y = \tau\zeta$; and for ζ we shall take $C_2 pv$, where C_2 is a constant, and v is the same as in Art. 103. And from the second point to B we shall use the transformation $\delta y = tz$; and for z we shall take $C_3 p$, where C_3 is a constant. These transformations are legitimate; for p does not vanish, except at D; and the points between P and Q may be so taken that v does not vanish between them. Let ξ_1 denote the abscissa of the point taken between P and D, and let ξ_2 denote the abscissa of the point taken between D and Q; then the term of the second order in the variation consists of parts under the integral sign, which are positive, since Q is positive, together with

$$\frac{1}{2}\left\{Q\left(\frac{z'}{z}-\frac{\zeta'}{\zeta}\right)(\delta y)^2\right\}_{x=\xi_1}-\frac{1}{2}\left\{Q\left(\frac{z'}{z}-\frac{\zeta'}{\zeta}\right)(\delta y)^2\right\}_{x=\xi_2}.$$

Now $$\frac{z'}{z} = \frac{q}{p},$$

$$\frac{\zeta'}{\zeta} = \frac{\frac{d}{dx}(pv)}{pv} = \frac{q}{p} + \frac{1}{v}\frac{dv}{dx}$$

$$= \frac{q}{p} - \frac{(1+p^2)^{\frac{3}{2}}}{2ayvp^2};$$

therefore $$\frac{z'}{z} - \frac{\zeta'}{\zeta} = \frac{(1+p^2)^{\frac{3}{2}}}{2ayvp^2}.$$

Thus, using the suffix 1 to apply to a value when $x = \xi_1$, and the suffix 2 to apply to a value when $x = \xi_2$, we find that the term of the second order in the variation

$$= \frac{1}{4a}\left\{Q\frac{(1+p^2)^{\frac{3}{2}}}{yvp^2}(\delta y)^2\right\}_1 - \frac{1}{4a}\left\{Q\frac{(1+p^2)^{\frac{3}{2}}}{yvp^2}(\delta y)^2\right\}_2$$

$$+ \frac{1}{2}\int Q\left(\delta p - \frac{\eta'}{\eta}\delta y\right)^2 dx,$$

where the integral extends over the whole range; from $x = \xi_1$ to $x = \xi_2$ we take $\eta = pv$, and for the remainder of the integral we take $\eta = p$.

The part which is under the integral sign is positive; but the other part is negative; for we have in fact implied that v_1 is negative and that v_2 is positive. Thus we cannot assert that the whole expression is always positive. Nevertheless, it is obvious that we might in many cases conclude that the expression is positive.

We know that δy must vanish and change sign in the course of the integration; see Art. 103. If δy vanishes twice within the range which includes the values v_1 and v_2, we may suppose these values to correspond to the cases; so that $\delta y_1 = 0$, and $\delta y_2 = 0$; and then the expression is necessarily positive.

[We may give another illustration of the process. Suppose that as p passes from its greatest value to zero v diminishes from the positive value λ to zero, then becomes negative and passes from zero to negative infinity; when p changes sign v changes sign also; see Art. 103.

Let $-\mu$ denote a negative value of v, and suppose μ numerically greater than λ.

Let us separate the integral into two parts, one extending from the initial value of x up to the value which makes v equal to $-\mu$, and the other from this value of x to the final value.

In the first part of the integral we use the transformation $\delta y = tz$, where $z = C_1 p\,(1 + mv)$; and in the second part of the integral we use the transformation $\delta y = \tau\zeta$, where $\zeta = C_2 p\,(1 + nv)$.

It will be possible to give such values to the constants m and n, as to ensure that $1 + mv$ does not vanish within the first part of the integral, and that $1 + nv$ does not vanish within the second part of the integral. For these conditions will be satisfied if we take m positive and less than $\frac{1}{\mu}$, and n positive and between $\frac{1}{\mu}$ and $\frac{1}{\lambda}$.

Then
$$\frac{z'}{z} - \frac{\zeta'}{\zeta} = \left(\frac{m}{1 + mv} - \frac{n}{1 + nv}\right)\frac{dv}{dx}$$
$$= \frac{n - m}{(1 + mv)(1 + nv)}\,\frac{(1 + p^2)^{\frac{3}{2}}}{2ayp^2};$$

and we find that the term of the second order in the variation becomes

$$\frac{n - m}{4a}\,\frac{Q\,(1 + p^2)^{\frac{3}{2}}\,(\delta y)^2}{(1 + mv)(1 + nv)\,yp^2} + \frac{1}{2}\int Q\left(\delta p - \frac{\eta'}{\eta}\,\delta y\right)^2 dx,$$

where the term free from the integral sign is to have the value corresponding to the value $-\mu$ of v. In the term under the integral sign we take $\eta = p\,(1 + mv)$ for the first part of the integral, and $\eta = p\,(1 + nv)$ for the second part.

The part which is outside the integral sign is negative. For $n - m$ is positive, $1 - m\mu$ is positive, and $1 - n\mu$ is negative.

We see then that if δy vanishes for any point for which the value of v is numerically greater than λ we may take $-\mu$ to correspond to this point; and then the term of the second order in the variation reduces to the positive part which is under the integral sign.

The value of Q is $\frac{y}{(1+p^2)^{\frac{3}{2}}}$ in the present problem; and by putting this value for Q some of the expressions will be simplified.]

CAMBRIDGE: PRINTED BY C. J. CLAY, M.A. AT THE UNIVERSITY PRESS.

PREPARING FOR PUBLICATION.

A HISTORY

OF THE

MATHEMATICAL THEORIES OF ATTRACTION

AND OF THE

FIGURE OF THE EARTH

FROM THE TIME OF NEWTON TO THAT OF LAPLACE.

Printed in Dunstable, United Kingdom

91220025R00167